Solutions to Problems in the Textile and Garment Industry

Solutions to Problems in the Textile and Garment Industry

B. Purushothama

WOODHEAD PUBLISHING INDIA PVT LTD

New Delhi, India

Published by Woodhead Publishing India Pvt. Ltd.
Woodhead Publishing India Pvt. Ltd., 303, Vardaan House, 7/28, Ansari Road, Daryaganj, New Delhi - 110002, India
www.woodheadpublishingindia.com

First published 2015, Woodhead Publishing India Pvt. Ltd.
© Woodhead Publishing India Pvt. Ltd., 2015

Woodhead Publishing India Pvt. Ltd. ISBN: 978-9-38030-849-4
Woodhead Publishing Ltd. e-ISBN: 978-9-38030-896-8

Typeset by Third EyeQ Technologies Pvt Ltd, New Delhi
Printed and bound by Replika Press Pvt. Ltd.

Contents

Contents

The textile industry which also includes apparel making is the oldest in the world, and is spread world over. Whenever a conversation comes up about the industrialization of any area, the first industry that comes to mind is textiles. Almost all industrial giants started with textiles, and then diversified to other industries.

Textile industry is not only the oldest, but also highly complex and competitive. The varieties produced are innumerable, and so are the problems. The industry being the oldest should have solved all the problems by now, but it has not happened. The industry is living with few problems which are self-created. The problems are repeated because people try to take the same steps which look attractive and money earning, without realizing that these are the roots of the problems. It is not that they do not know, but they try to take a chance; just like driving on the wrong side of the road to reach early.

The textile and garment industry faces problems not only because of internal issues and competition in the market but also due to policies of the government, the world economy, political issues, developments in other industries like IT, automobile, and infrastructure, and so on. In this book, different types of problems and problem solving techniques are explained with different case studies. It is not possible or practicable to give readymade solutions to the problems therefore, different techniques used at different times are explained so that the reader can think and adapt appropriate technique depending on the situation and gravity of the problem. If readers can explore different ways and means to solve their problems, I shall be happy as a writer of this book.

I hope this book shall help the industry.

B. Purushothama

Foreword

There is no person or organization that does not without a problem. As the problems are solved or overcome, one reaches next height. Textile and apparel industry are not an exception.

All problems are not same. Even for similar problems, same solution that was successful at some place might not be acceptable due to a different situation and environment. Hence, one has to apply his mind depending on the situation. In this book, different types of problems and the techniques which could be applied to those problems are explained along with case studies related to textile and apparel industry. This book can be used for other professions also.

This book is first of its kind in the world that deals with various practical problems faced by textile and apparel technicians and management, and stimulates the mind to adapt and also develop different techniques for solving their problems.

This book can be used by all, including technicians, non-technicians, professionals, consultants, or any person interested in solving his or others' problems, may be in textile industry or anywhere on the earth.

In the wonderland of problems

1.1 Textile industry

Textile industry is the oldest industry in the world, which is spread world over. According to DNA studies of clothing line, the clothing may possibly have been used 650 thousand years ago for the first time. A number of developments have taken place based on the practices followed in textile industry. In fact, the Industrial Revolution started with the Spinning Jenny developed by James Hargreaves. It is the reference model for developing the labor laws, factory acts, social accountability standards, quality management systems criteria, labor welfare measures, social insurance systems, and safety standards, and so on.

Whenever there is a conversation about the industrialization of any area, the first industry that comes to mind is the textile industry. Almost all industrial giants in the country started with textiles and then diversified to other industries.

India being the cradle of textile industry, the concepts of growing and cultivating cotton and producing fabrics here were in practice since 4000 BC. Skill in the manufacturing of cotton was highly developed, and was handed down from one generation to another. The spinners and weavers of India, in spite of their excellence in skill, had no organized bodies like the guilds of Europe. On the contrary, carders, spinners, weavers, dyers, printers, etc., were segmented into sub-castes. Their income was low and their future depended on the goodwill of merchants. The expertise they achieved was due to the refined tastes and love for luxury of the rich people; this ensured the prosperity of the textile industry. In urban areas, the industry was organized in form of *"Karkhanas"* that thrived on the royal patronage. In the reign of Aurangzeb, Francois Bernier observed that a wide variety of piece goods were being produced by these *"Karkhanas"*. But in the country as a whole, the cotton textile industry was essentially a family enterprise and depended on a large and assured home market. Economically, it was second to agriculture.

India is the world's second largest producer of textiles and garments after China. Also, India is the world's third largest producer of cotton (after China and the USA) and the second largest cotton consumer after China. The Indian textile industry is as diverse and complex as the country itself and it combines this immense diversity with equal equanimity into a cohesive whole. The fundamental strength of this industry flows from its strong production base of wide range of fibers and yarns from natural fibers like cotton, jute, silk, and wool to synthetic and man-made fibers like polyester, viscose, nylon, and acrylic. The growth pattern of the Indian textile industry in the last decade has been considerably more than the previous decades, primarily on account of liberalization of trade and economic policies initiated by the Government in the 1990s. The textile industry moved from producer-driven value chain to a buyer-driven value chain. In producer-driven value chains, large and usually transnational manufacturers play the central roles in coordinating production networks. This is typical of capital and technology-intensive industries. Buyer-driven value chains are those in which large retailers, marketers, and branded manufacturers play the pivotal roles in setting up decentralized production networks in a variety of exporting countries, typically located in the developing countries. This pattern of trade-led industrialization has become common in labor-intensive consumer-goods industries such as garments and handicrafts. Large manufacturers control the producer-driven value chains at the point of production, while marketers and merchandisers exercise the main leverage in buyer-driven value chains at the design and retail stages. The relative ease of setting up clothing companies, coupled with the prevalence of developed country protectionism in this sector, has led to an unparalleled diversity of garment exporters in the third world.

Textile industry is not only the oldest, but also highly complex and competitive industry. The varieties produced are innumerable and also have its problems. The industry being the oldest should have solved all the problems by now, but it has not happened. The industry is living with few problems that are self-created. The problems are being repeated because people try to take the same steps that look attractive and money earning, without realizing that they were the roots of the problems. It is not that they do not know, but they try to take a chance; just like driving in a wrong side of the road to reach early.

The textile and garment industry face problems not only because of internal issues and competition in the market, but also due to policies of the government, the world economy, political issues, developments in other industries like IT, automobile, and infrastructure, and so on. The textile and garment sector, especially garments, is one of the most globalised of any in the world economy. The world recession has hit the Asian textile and garment sector at a time when the sector globally is struggling with potentially massive readjustment (UNCTAD 2005). John Thoburn observes that globalization, however, owes much less to normal market forces than to trade distortions, particularly the Multi-fiber Arrangement (MFA) and its successor, the Agreement on Textiles and Clothing (ATC). The MFA/ATC, which ended on 1 January 2005, controlled exports to major markets, particularly the EU and the United States, for over thirty years. Garment, typically high labor-intensive and less capital intensive activity, compared to textiles helped developing countries to enter world markets as manufacturers and exporters. In the case of garments, such entry has been greatly aided by the MFA's restrictions on the most competitive countries like China.

Within the framework of the MFA, rising wages have been an important driver of international relocation in garments and to a lesser extent in textiles. Early established producers of textiles and garments, driven by rising wage costs and the cost and availability of factory sites were motivated to retain their competitive advantages in the international economy by shifting their production to lower wage countries. Within India, a number of industries moved out of metro cities like Mumbai, Bangalore, Delhi, Chennai, and Ahmadabad and were relocated in rural areas. Of course, the process of cost-driven relocation was not a wholly Asian phenomenon but was a global phenomenon.

Textiles have had successive waves of innovation, often resulting from innovations in other industries, in particular chemicals, machinery, and information technology. This has increased the industry's capital intensity and made textiles less prone to relocation to lower wage countries. In garments, although few innovations took place in CAD and in cutting, the major activity remained labor intensive and the wages played an important role. In addition, the employment of nearly 80% female workers restricted some of the freedom which a textile mill has like round the clock working, fixed employees attending to work on regular basis, etc. The garment factories are facing the problem of too many attritions of female workers, restricted hours of working, and so on.

Global-value chains refer to the process whereby the successive economic links in the production process are organized not by arms-length market transactions but by longer-term contractual arrangements in which economic power is concentrated in the hands of economic actors at particular stages of the chain. In the case of textiles and garments where power is exercised at the retail end, the value chains are buyer-driven. Although rising wages have been a driving force of location change for labor intensive industries like garments, and to a lesser extent for textiles, the control or governance of such relocation has been primarily in the hands of global buyers.

Prior to the end of the MFA, the pattern of buyers' international sourcing was driven by required lead times for different kinds of garments, quality of workmanship and price, MFA quota costs and availability, and import duty payable in major markets on exports from a particular location. These requirements differed between ultimate buyers according to the market segment they serve and are the basis for the subdividing of buyers in the existing literature. For example fast fashion items in ladies wear require a short lead time, while more traditional items like men's dress shirts or suits less so. Basic items like T-shirts or denim jeans also do not require short lead times and are likely to be sourced on the basis of the lowest costs. The requirements of particular final buyers themselves also are variegated according to the range of goods they wish to retail. Short lead times favor nearby suppliers whereas for cheaper basic products buyers can go further afield to places.

The forces driving changes in global-value chains have continued into the recession. Intense competition in the United States and European retail markets has made for greater concentration of sales. Buyers have attempted continually to lower their buying prices while maintaining quality and striving for shorter lead times to meet new fashion trends. The United States market remains somewhat different from that of the European market. The United States buyers tend to source United States country-wide, with very large orders, great sensitivity to price (especially by buyers at the lower end of the market), and to be less loyal to suppliers over time. Hence, any amount of hard work done by garment industries is not paying them when they are catering to such big buyers. The European markets are more variegated in different countries; but in general order as compared to the United States, they tend to be smaller, quality requirements higher and loyalty to suppliers is greater. The Japanese market resembles the European market, except that the orders are still smaller, the requirements for quality even higher, and there is less willingness to switch suppliers since the Japanese buyers take trouble to develop their suppliers' capabilities. Many garment factories have failed mainly because they chose a wrong customer who only demands but has no loyalty towards the suppliers, and does not provide any support.

Irrespective of global-value chain problems and the government policies, normal problems faced by the industry are high employee absenteeism and attrition, very high competition and low profit margins, too much fluctuations in raw material prices, ever increasing cost of manufacturing, volatile market conditions, and ever changing policies of the government and trade. In this book, an attempt is made to compile various techniques available world over and to verify how they can be used in textile and garment industries to solve their problems.

Before discussing on the problems and remedies for the industry, let us try to understand the basic meaning of a problem, different methods available for solving a problem, and the method of choosing the best alternative.

1.2 What is a problem?

We always talk about problems. There is no one without a problem. But all might not be having the same problems. Something is a problem to one, but the same is not a problem to the others. Similarly, their problem is not my problem. If one says, "corruption is a problem for development of a nation", the corrupt person would say that "the honest people are the cause of the problem". He would say that "the honest man is always very strict, and does not pass any bill or resolution as he finds one or the other loop holes; whereas the corrupt person would accept give something to ensure that the work is done so that we move forward, hence improvement". If lack of knowledge is a problem in a number of cases, having extra knowledge could also become a major problem or threat in few cases. For lack of knowledge, a man can be educated or trained, but what will you do if a man has knowledge about the malpractices being done by you the top management or bosses in the department.

Then what is a problem? There are different definitions to it. According to Concise Oxford Dictionary, it is a doubtful or difficult matter requiring a solution. It is also defined as something hard to understand or to deal with. A problem is defined as an undesirable result of a job, as per the Japanese concept of Total Quality Management. For example, when a thief is gently opening the safe and something falls down making a noise. Or when you are sitting with your girlfriend in a garden, your wife suddenly comes there.

The solution of a problem is to improve the poor result to a reasonable level. If we solve our problem, we can move forward. If we leave it, it shall go on growing. That is the reason the elders say, "to be a winner, one should stand firm and face the problems". Also they say, "One who is afraid of a problem can never win". So problem solving technique is a starting tool in the modern management techniques. But if we see different problems, one might feel that there is no need to solve all our problems, and sometimes we cannot even solve the problem. In such cases, it might be worthwhile to live with the problem or run away from the problem.

There are few people who do not like their problems to be solved, so that they can are by that. Few may win the elections, and few others might get some compensation on a regular basis. Few people build a hut near the river, although they are aware of the floods. Every year the hut is washed off and they get compensation for it. Another classical example is the disputes relating to sharing of water between states. Until the problem is not solved, few politicians shall be getting their votes. Sometimes those who are paid for solving the problem try to pull it for a long time so that their income is not affected. Take as an example of prolonged court procedures even for simple items, and there shall be number of adjournments. If one goes to court, it would be adjourned due to some reason. There are cases of land disputes, language disputes, reservations, etc., that remained unresolved for centuries. If one tries to solve them, some other puts an obstruction. In industrial life also the same is observed. Few industries want them to be declared as bankrupt so that they can sell off their lands and make money for self but do not have to give whatever is due to the workers, suppliers, and to the community (taxes).

There are certain categories of people who always see a problem in anything they see or encounter. If you try to help them, they suspect you of getting some benefit. They suspect everyone.

Sue Dinwiddie observed that in their very early years children develop their coping skills for how to handle conflict: fighting back, running away from the conflict, or problem-solving, a win-win solution. For centuries adults handled conflicts with young children by telling them what *not* to do. Sometimes these words were reinforced with punishments to increase compliance. This was a negative process for both children and adults, often ending in revenge. With proper guidance, children recognize that they have choices and can find an acceptable way to achieve their desires.

1.3   Hill as a problem

In Figure 1.1 we can see a hill, which was obstructing the road. People had to take a long route because of the hill. The time and fuel consumption was high. There was a demand for improving the road and therefore, the government decided to construct a tunnel. Once the tunnel was made, the problem was solved. Vehicles now could easily pass through, saving time and fuel. Whatever the money spent for the construction of tunnel could be recovered easily within 4 years because of the savings in fuels directly. Time saved is difficult to measure in terms of money. This is an example of solving a problem that lead to improvement.

While trying to understand the benefits, the perception of the people is an important point to be considered. Whether the majority and the people living nearby wanted this road or not, and why they wanted the road there? Was there no alternative? Few protest because of loss of their lands and few others for their prestige. An environmentalist makes a hue and cry of damage to ecology and may arrange for protests and demonstrations, and few miscreants enjoy by burning few buses in some other city to save this hill in an interior part of a forest. A politician from opposition talks of misappropriation of funds, and not giving preference to some other area which was more deserving. Each problem should therefore be viewed from different angles either as an opportunity for improvement, a threat for survival, a hurdle for progress, or something hurting our feeling or ego. Whatever may be the point of view, normally people prefer taking suitable actions to overcome the problem.

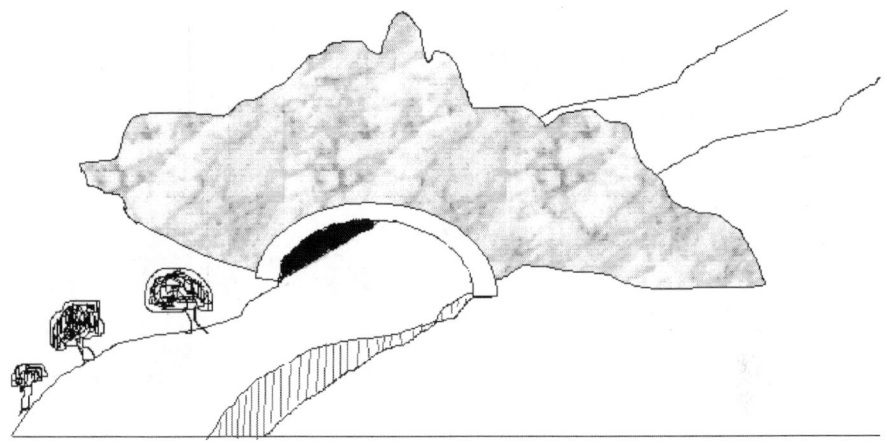

Figure 1.1   Hill – Problem in a road.

In textile industry, permissions were not given to modernize weaving in the mill sector up to 1989 in order to protect the interests of decentralized power-loom weavers and handloom weavers. This was termed as the major reason for mills becoming sick; whereas after the permissions were given to modernize the loom shed in mill sector, still the mills were in loss. Whether lack of modernization was the reason for losses or the lack of competency to run the mill was a problem is the million dollar question remaining unanswered.

1.4   Hill as a beauty spot

If something is a problem or not, depends on the person seeing it. If people enjoy going round the hill while seeing the beautiful scenery, then the hill is not at all a problem but an added attraction for this road. One could encash by developing tourism in that area. People would demand it being preserved.

If one can understand the specialty of slow-speed old machines, they can successfully utilize them in producing novelty products that cannot be produced even on high speed machines. The people who purchase old machines from mills are running them and making profit, whereas the mill which sold the old machine and installed brand new machine of latest technology is in loss.

Figure 1.2   Hill as a beauty spot.

Life is a combination of failures and successes. The problem, if not solved shall result in a failure. Success can be achieved by overcoming or solving a problem. Sometimes avoiding a problem might be the best solution or adjusting self to the problem. There are few problems which cannot be solved, but we can adjust to them. For example, in winter cold is a problem. One cannot reduce the cold, but can wear a sweater and protect oneself. The birds from Siberia fly all the way to South India to prevent themselves from cold and again fly back when the weather is suitable. They know how to avoid the problem and do not wish to live with it or face and fight it. When a lion roars, all the deer run away to safety. They do not think of standing firm and facing the problem, or they do not try to device any plan to solve the problem. They chose only one option and that is to run away from the problem. That is the only solution for this problem. The mills who adapt to the changing situations and divert their products and services to suit the needs of customers shall survive, whereas the mills sticking to a particular product or market, or a particular pattern of working shall find it difficult to survive during competition.

It is the nature of a human being to take advantage of whatever is available, just like viewing a hill as a beauty spot rather than an obstruction for movement. In the mills where the workers have to work for 12 h with no weekly off and also with no rest room facility find their own ways. They happily sleep wherever they feel comfortable as shown in the Fig. 1.3.

Figure 1.3   Sleeping in the department.

There are few cases where we do not like either to run away from a problem or to face it and fight, but like to surrender and win. For example, your spouse wants a ring or a bangle for the marriage anniversary and you do not have enough money. Her demand is a problem, but you cannot convince her. By opposing, the problem is going to increase and it is better to surrender and somehow adjust. You might sacrifice some of your interest, but win the heart of your spouse with whom you have to live for life time.

In Indian culture, the marriage is a life time bonding. Whatever may be the problem, the couple does not get separated. They somehow learn to adjust and live together. They face all problems, take all sufferings, but never think of separating. So the family remains as one, and the children come up. The act of divorce is an act of cowards who do not have the guts to face the problem and win over the situation. They make the children orphan although the parents are alive. They are deprived of the love and affection of parents. In Indian system, the quarrel between husband and wife is often compressed by the love and affection they have for their children.

The textile mills in India normally adopt this formula of "pleasing the spouse", i.e. the officials of the local government who object for non-implementation of rules and regulation are pleased by offering something and made to cooperate just like a husband pleases and enjoys with the wife.

The officials have found an easy way to overcome few of the "*problems*" being faced by their industry friends like adhering to the Factory Act, paying minimum wages to the employees, providing leave facilities, providing double wages for overtime, providing maternity leaves to ladies, maintaining crèches, depositing the provident fund collected in time, paying taxes, and so on. Few call this technique as "*successful management*" whereas others grumble and brand it as unethical, scandal, and so on. The industry refers to this system as "*Normal Industry Practices*". The Press people, if they do not get their share, publish it as "scandal", otherwise demand their share in order to cooperate in the great activities being done.

Sometimes ignoring a problem is the best solution. Normally, the government officials and quality management auditors do this. They know that the mill is not maintaining the working hours as per norms, not maintaining the safety systems, not providing the leave benefits to their employees as needed, and not following a number of rules specified by the government, but they totally ignore it. Even the auditors sent by the buyers to ensure implementation of social accountability standards also ignore the violations, because if they object and disqualify the company, they will not get the materials at the rate provided by these people. The buyers want materials at lowest rates but also want to show to the society that they are purchasing materials from the companies that are ethical, following all regulations of government and social accountability because of which they are accredited with ISO 9000 and SA 8000 certificates.

Now a days, we are seeing advertisements on internet that "Pay Rs 9999/- or Rs 10500/- and get ISO 9001 or SA8000 or ISO 14001 certificate in 5–7 days without undergoing any formalities of audits". Although these companies are advertising openly, the authorities like Quality Council of India or Bureau of Indian Standards are ignoring them. If they have to take legal action, they have to spend money and also become "bad" in the eyes of few of their well-wishers. The institutes that are supposed to monitor the implementation of ISO guidelines have become "Real ISO" companies, i.e. "*Intentionally Sleeping Organizations*".

ISO 9001-2008 Certificate
Certify your organization's qualitymngt STD, by ISO 9001 2008.Visit us
www.nimbuscertifications.com/9892413039

ISO 9001:2008 in Rs 10, 500
for 3yrs, Easy&Fast Cert in 5 days. Just Call: 09350942153 / 7428669150
www.iqmindia.com/9001.html

ISO 9001 @ Just Rs 9999
Get ISO Certification in 7 Days.Fast & Easy. Call: +91 8105999555
www.isocatalyst.com/register-now
Ads

Figure 1.4 Advertisements for getting quick certificates.

You have few ideas and are trying to implement them. You are sure of the results, whereas others are not. They discourage you by telling all the bad effects of your idea. Sometimes people may protest and write articles in papers against your process or project. Sometimes they may group people to protest and work against your ideas. If you are strong, they have no power to stop you. You are like an elephant walking on the main road and others are like dogs barking. The elephant need not bother about the barking dogs. Once the elephant crosses that road, the dogs stop barking. People get tired and stop opposing. Once you get the results and prove yourself as successful, the same people who were opposing you might come forward and congratulate you in front of all and claim that they were all your partners in success and you got the success because of their support and guidance. Ignore that as well.

There are instances when a problem was solved or overcome by creating another problem. Take an example of a minister involved in a scandal about whom the press has published. There shall be a hue and cry, and few protests. The opposition starts demanding the expulsion of that minister. Suddenly on a fine morning, someone sees that few miscreants have defaced the statue of a leader of certain faith. People start protesting against that and forget the scandal of the minister. There shall be a difficult situation for the government to control the situation. Suddenly a statement comes from someone that water cannot be released to the neighboring state, whatever might be the consequences. The attention would divert on that side. Then some controversial person is proposed for an award. Like this, they go on creating one or the other issues to ensure that people forget the original issue and the minister can live peacefully.

We can see such types of problems getting created to divert the attention of people from the present problem in textile and garment industry as well. When a supervisor is tackling with his tailors for the low production in a garment factory, someone will raise an issue of misbehaving or using abusing language with the girl by the supervisor. When the quality problem is highlighted by quality checker and the production person is ignoring it, the attention is diverted with an issue of late reporting resulting in shipping delays. When a worker is confronted for his poor quality of work, not providing leave and unavailability of rest room come out as issues.

Postponing the issue is one of the tactics used by many leaders. They resolve to form a committee, who meet many times, discuss the subject in detail, and give a report consisting of 2000 pages. The report shall not be read by anyone as no one is having time to read it. A few headlines shall be published in papers, and someone would oppose it. The issue shall be kept alive for years, and sometimes for decades and centuries. People will almost forget the issue and shall be doing their work, but whenever the leaders are in trouble due to some other issues, this problem shall be brought to the surface, and attention of people would be diverted. Some other committee shall be formed to study the issue and give report, which shall again take few years. One after the other committees shall be getting formed, and different views shall be coming from each committee. There shall be no solution, as the leaders would not want to solve the issue. They can exist only if a problem exists. In textile industry, the group incentive schemes, workload settlement, and wage fixations are such issues. No conclusion is drawn and the problems are kept pending so long that one day people forget it.

Sometimes sincere efforts are made to solve a problem, but they end up with another problem.

(a) A dam constructed to provide drinking water, irrigation, generating power etc., submerges huge forests and valleys, and few villages or township in the valley. Similarly, a good wide window constructed to provide good light and ventilation allows mosquitoes, dust, and noise inside. It creates huge amount of problem by affecting the working conditions in a cotton mill. A false ceiling provided in a spinning shed to reduce humidification expenses becomes a potential fire hazard.

(b) To encourage a sincere employee, promotion was given to him that made others unhappy. An incentive plan implemented at winding department made workers of warping department unhappy. The wage revisions made in wet processing made spinning people unhappy, who went on a strike.

(c) To avoid undue delay in launching of a project, bribe was given to some official and it became a practice. Now the demands have increased and one cannot afford that much money to give and workers are suffering. Mill is not sure of making money even with the launching of new project because of competition. The government officials are demanding bribe even to run the present system as it is.

(d) I took medicine from a doctor for some disease, and that resulted in another problem as a side effect. I introduced attendance bonus for the doffer boys in spinning as the absenteeism was very high, but the people in processing also started demanding attendance bonus as they were very regular in their work.

(e) Pesticides used to control pests in a farmland killed the pets also. The decision taken to delete old unwanted files from the computer system also deleted large amount of useful data.

(f) Mercy was given to a culprit considering his family conditions, and now he demands it as a right for all his misdeeds.

(g) The implementation of computer-aided information systems reduced the thinking ability of the staff working in textile and garment industries.

(h) Annual maintenance contracts given to machinery suppliers to help good maintenance made the company technicians dependents, without any initiative and knowledge for identifying and solving the problems.

1.5   Road with and without hurdles

Normally no one wants a problem, but wants to grow and be an achiever. It is not possible all the time. One has to face his problems and overcome them in order to achieve anything. Whatever you get without struggle or efforts is not an achievement. Even if you do not want to achieve anything and stay where you are, you need to do some effort.

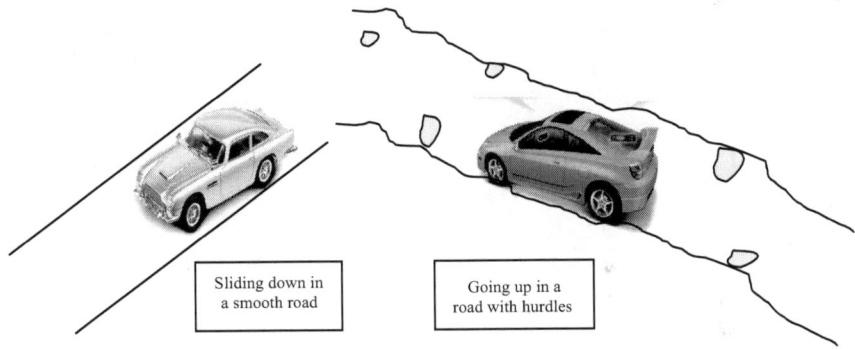

Figure 1.5   Road with and without hurdles.

If you like to go on a path where you feel there are no problems, it means you have taken the easiest route on which all are moving. In such a case, you are one among them and cannot be a winner. Don't be very happy if you are moving at a high speed without any hurdles. Be careful; the road may be leading to a ditch. Remember, a road without hurdles can take you down and not up. When you have to go up, you have to struggle, make efforts, and might come across many hurdles. You should face them. You have to overcome them. One who is afraid of facing problems can never be a winner. He can only have a temporary safe shelter like the deer in the forest. But as a human being, that too when you want to be an achiever, be ready to face the problems. Do not try to avoid them or postpone your actions when you have capability to face it and win. If you are hesitating to move forward, someone shall overtake you and you will not come out as a winner. Not moving towards the required direction in time is your problem.

In production of a textile mill, less changes means a smooth working environment for the technicians. Frequent changes in product mix like changes of counts and blends in yarns, changes of designs in weaving, and changes of processes and finishes in wet processing, and changes in styles in garments leads to stoppages, quality problems, problem of production balancing, problems in fixing targets for production, problems in planning and maintaining of machinery, problems in timely procurement of required raw materials and accessories, and so on. Any technician would prefer long orders and less changes, whereas the management would want more changes as the profit margins are high in smaller lots rather than in bulks. Too many changes bring down the efficiency and morale among the people working, but the management claims it to be the strength of the company. So the people working in fashion industry are always on their toes and shall be daily working extra hours without any rest and in majority of cases, without any compensation for the extra hours worked. They get fed up and leave the company, but get a job worse than the present and the struggle continues.

Any job will bring problems to be faced. It is important to show the recruiter that you have the right skills to resolve these problems and the personal resilience to handle the challenges and pressure they may bring. You need to be able to:

(1) Evaluate information or situations

(2) Break them down into their key components

(3) Consider various ways of approaching and resolving them

(4) Decide on the most appropriate of these ways. Always remember that Problems can also be opportunities: they allow you to see things differently and to do things in a different way; perhaps to make a fresh start.

Whatever issue you may be facing, few steps are fundamental as follows:

• **I**dentify the problem.

• **D**efine the problem.

• **E**xamine the options.

• **A**ct on a plan.

• **L**ook at the consequences.

This is the **IDEAL** model of problem-solving.

The innovation can solve a number of problems. While it has been successfully used many times to improve society, many societies do not encourage it. Instead, they encourage people to follow the crowd. This is hundred percent true with many of the textile mills in the country.

1.5.1 Wicked Problems

Many problems or puzzles can be solved with a number of different methods such as trial and error, brainstorming, and reductionism. These problems could be called static or well defined. Few problems cannot be solved fully, or have requirements that continue to change. This type of problem is much more difficult to solve and is referred to as being a "wicked problem." The idea of wicked problems was first developed by H. J. Rittel and M. Webber. The solutions to wicked problems are hard to solve because the elements that compose them continue to change. When one attempts to solve a wicked problem, he may find that the solution is creating yet another problem.

Any problem that requires a large number of people to solve is also called a wicked problem. The reason why these types of problems are hard to solve is because they will require a large segment of the population to change their views, and this is extremely challenging. While solving a wicked problem may be next to impossible, Rittel did create a system that would make the problems easier to manage. This system is called IBIS, or Issues Based Information System. It will allow large number of people to break down problems into both questions and arguments.

The first step in solving a wicked problem is to structure and analyze it. The proper name for this is morphological analysis. There are four rules that are associated with a wicked problem. First, the problem will not be well defined until a solution has been created. The second rule is that stakeholders will have different views when it comes to understanding the problem. The third rule is that the resources and barriers involved with solving the problem will continue to change as time goes on. The fourth rule is that the problem can never be completely solved. There are no unlimited solutions with wicked problems, and every wicked problem will be connected to another problem. Any lessons that are learned from a wicked problem cannot be used to solve other wicked problems because each wicked problem is different.

The only way to come close to solving a wicked problem is to have a large group work together. In a sense, if a large group of people can get together and brainstorm, this may allow few relevant solutions to be developed. However, these problems can only be solved with creative methods. Logic is not as useful in solving wicked problems as it is in solving other problem types. The biggest disadvantage to come up with a solution to a wicked problem is that the solution will always have consequences, and these consequences will lead to another problem. Wicked problems are evidence that problems are designed to exist. There is no way to solve them completely.

1.6   This is your mistake

"I want to do something, but others do not like it. Sometimes even though there is no problem, people make comments. I am really fed up with the comments..." You might have heard people saying this. Few people say, "In some office, the bribe is to be taken because other colleagues are taking and if I do not take, I shall not get cooperation from my workers; remember, I have to give if I need to get my work done."

Management recruits a senior manager working in a reputed company as their Chief Executive with a hope that all good systems of that reputed company could be implemented in this company also, but unfortunately, the new CEO feels that he should work as per the other people in the organization, and shall not bring any change. He defends himself by saying the famous proverb "Be a Roman when you are in Rome".

Few of the mill managements are always trying to hunt good managers from reputed mills by luring them with high packages, but normally the people falling in their net are the ones who are a failure in a reputed company and are searching for an opening elsewhere as he is not sure of continuing there. Good managers do not leave good companies as the companies take all the care to retain them.

Few mill managements claim it as their intelligence of luring and taking staff from other good companies by paying them handful salaries as they are avoiding the expenses of training and trials, whereas they are unable to work out the losses incurred in losing the well-trained staff and workers who knew the systems and were fluent in handling the company situation. The person whom you forced to leave your company joins another company from where you have drawn a person and there was a vacancy. Both the people get higher salary than what they were getting but shall be doing the same work what they were doing.

We need to think and work out what is needed for us, and how to overcome our problems. If one goes on listening to other's opinion for all his deeds, everything can be described as a mistake only. Few of the examples are as follows:

(1) You are very careful and do not want a mistake to happen. You are taking all care to see that everything is perfect. This is testing the patience of others. They are not happy with you as you seems to be slow and over cautious. They say you are not practical. This is your mistake. In textile and garment industry, this can be seen in almost all companies. Management wants quick results by hook or crook, and the people who are methodical are branded as theory persons and not practical.

(2)   You did not want a mistake to happen. Hence you studied the system and documented all the steps. You decided the control and check points, and suggested all to follow it. Others did not want any such things. They say you are not a practical man. This is your mistake. This can be seen commonly in textile and garment industry.

(3)   You did not want a mistake to happen. Hence you studied the system and documented all the steps. You decided the control and check points, and suggested all to follow them. Others could not understand your system fully as it was not competent. Everything failed. This was your mistake. This happens in majority of the mills in India as people are employed without referring to the minimum competency required for the job being handled. The latest machines with PLC controls and more automation require people who are more methodic.

(4)   Someone is doing something wrong, and you suggested him not to do it. He did not like this. This is your mistake. Similarly, a man has been brought by a man with influence in the top. He may be a relative of the chairman or managing director. If you point out his mistake, you shall be out of the company.

(5)   Someone is doing something wrong and you saw it, but did not do anything to prevent it. Wrong things happened in front of your own eyes. This is your mistake. The top man questions you as to why you did not bring this to his notice. As the people close to the management are likely to be involved in malpractices, you have been appointed to control them. They have been appointed on some obligation, but it does not mean that they can siphon out the earnings of the company.

(6)   Few of your friends have a scheme by which they can earn fast. You are not confident or you feel that it is not ethical. Therefore, you did not join them. They did not like this. This was your mistake. Now they will start making a team to work against you, as you might be a threat to them.

(7)   Few of your friends had a scheme by which they could earn fast. They wanted you to join them and you joined them. You were caught and it was your mistake. The management wanted you to be ethical and honest and dedicate your time for the well-being of the company and not for involving in any schemes for personal benefits.

(8)   Someone did a good work according to you and you appreciated him in front of another person. The second person did not like you appreciating him. This was you mistake. By doing this, you are branded as a man without maturity and cannot understand the politics.

(9)   Someone did a good work according to you and you did not appreciate that. The one who did the good work lost interest and stopped doing

good work. This was your mistake. You do not know how to motivate good workers to do better work.

(10) One of your assistants is very good at work and can do any type of work. You recommended him for promotion. Other seniors did not like your decision. This was your mistake. You are violating the company's policy and practice.

(11) Your lady assistant is very sincere and hard working. You appreciated her work in front of your wife. That was your mistake. You have to pay for it throughout your life.

(12) Your lady assistant was very sincere and hard working. You did not recognize her work. She got upset and left. This is your mistake. The company lost an efficient worker.

Thousands of such examples can be given from textile and garment industry. We need to understand what is our problem, what is required for us, whether we are moving towards achieving the goal or not, whether we can survive or not, etc., and have a determined approach while dealing with the problems. Let us move further to understand the roots of the problems being faced by individuals as well as the industry in this wonderland of problems.

2.1 The tree in my yard

People say, "Remove the problem by root". For that we should know where the roots of the problem are. What we normally see as a problem is not the real problem but is a symptom or an effect of a problem. Take the example of a tree in my yard, which I do not want. It is a problem for me as it sheds lots of leaves, and house is getting cracked because of its roots. The monkeys come and play on the tree as well on the tiles of my house, cracking the tiles. I have tried to cut the tree off, but still it is growing. I am not putting any water, but it is managing. It is taking water from the neighbors. It has grown very big. What can I do? How to solve the problem?

Roots of the problem are not in my compound

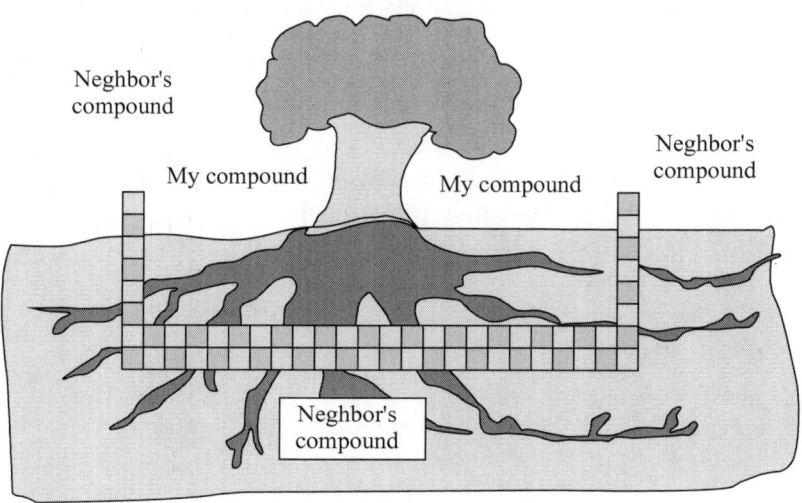

Figure 2.1 The tree in my yard.

When Dr Juran said that 80% problems were management related, people laughed at him and branded him as mad. Let us see an example; I have a tree that is a problem to me. Whether tree is the real problem, or the monkeys breaking tiles, or roots damaging the foundation? Who is responsible for that? I did not remove it when it was a small plant. I enjoyed when it was small and encouraged it to grow. This was my first mistake. The roots spread and I did not make an effort to clear it, but allowed it to grow. Now I am not putting any water in order to stop its growth, but it is getting water from somewhere else. Are my neighbors responsible for this? No. They are not aware that the roots have spread in their yard. Why did the problem arise? The land and soil in my yard are suitable for that tree to grow. My systems are favorable for the problem to grow. My lethargy or negligence of not taking any action in time supported the problem to grow. I should have reinforced my land by putting stones or cement, so that a plant cannot grow. I did not do anything to prevent a problem. Now what should I do? How can I come out of the problem? I should take the neighbors into confidence and seek their help. I should convince them that I am in trouble, and if they do not mind, they could help me. I should ensure that I am not harming any of the plants in their yard, and after taking their consent dig the roots out from their yard. Once all the roots are removed, I have to put stone pavement to prevent a second plant from growing in my yard. If I am unable to convince my neighbors, I will not be able to take their help. The problem shall remain and I shall have to learn to live with it. I cannot force my neighbors to allow me to dig in their area.

The real problem is not the tree or its roots. Neither my neighbors are problem creators, nor are they supporting the problem. It is my lethargy, negligence, and incompetence to convince my neighbors. The root cause is me.

This holds well for all the problems and for all organizations. Examples of textile mills are as follows:

(1) A mill owner brings few of his close relatives and friends and gives them responsible posts in the company as he does not trust others. The same people whom he trusted siphon out the funds without his knowledge. The problem is not the integrity of the people, but not devising a transparent management system and not having control on day-to-day activities irrespective of whether it is relatives working or any paid employee.

(2) Few workers were encouraged to spy over the activities of their bosses and to inform the chairman. After few days, after knowing the weaknesses of both chairman and the functional heads, these people started black mailing both of them to get personal benefits.

(3) When there was a boom period, the mill decided to expand. While installing the machines time was not wasted to make a master plan and the machines were erected at the place available, considering the production at that time. It became haphazard when the expansions were made, the material movement was not free, the places for stacking of materials were not sufficient, and finally the working was not efficient.

(4) Contractual workers were encouraged to reduce the cost of production and to get rid of the problems likely to be encountered if the workers were permanent; but these contractors vacated as they got better offers elsewhere and the mills had shortage of workers to work the mills and production was suffered.

(5) Managers were encouraged to bring workers with them when they were recruited to get sufficient people for the mills. Each manager was insisted to bring more workers and the one who brought more workers was given higher post. One day, that senior manager got a good offer and took all the workers he had brought along with him. The mill was completely deserted.

(6) A worker did good work and gave expected quality and production whereas others were struggling; a cash reward was given to him. Within a short time, all others learnt the system and everyone started giving the expected production and quality, and demanded management to reward all in the same way as that one worker.

(7) A mill which was normally running in loss made profit once and the management gave a bonus of 20% in place of normal 8.33%. Next year, the mill went in loss, but the union insisted on the same 20% bonus saying that making loss or profit was in the hands of management and not the workers.

(8) In a spinning mill engaged in exporting yarns ladies were appointed to remove contaminations from the cotton, and one bale was given to each lady. As the quantity of exports increased, the number of the ladies also increased as management was forced to accommodate them. When online foreign material detectors were introduced, the management faced a tough time in order to remove these lady workers.

(9) A strict instruction was given by the management to ensure that no employee comes to the factory with *gutka*, cigarette or *beedi*. This forced the security personnel to check all the incoming employees, because of which workers were detained at gate. There were no workers to start the work in the beginning of the shift and production had to suffer for 30 minutes.

(10) Refilling of the fire extinguishers was not done as the management did not feel the necessity. There were no cases of fire in that garment factory earlier, and the fire extinguishers purchased long back were all idle. It worked without any accident for 3 years. Gradually, people forgot about refilling the fire extinguishers and conducting firefighting drills. When there was a fire, none of the staff or workers was aware about how to fight the fire and therefore, the factory had huge loss.

2.2 Theory of 1-30-300

The analysis made by American Association for Safety and Health revealed that behind every fatal accident there were 30 near misses and 300 unsafe conditions. To prevent one fatal accident, one has to concentrate on proactively eliminating 300 unsafe conditions. 1-30-300 accident pyramid explains the concept in a simple way. Mr. Shyam Talwadekar explained the same with illustrations in his book "Visual Management" and stressed on the elimination of the problems at seed level itself by good work practices. He suggested using 5-S concepts to identify the problems in their seed stage and eliminating them. This holds good for the example of "Tree as a Problem" explained earlier. If I had removed the seeds, the plant would not have come up. If the action was taken in the initial stages, the problem would not have existed.

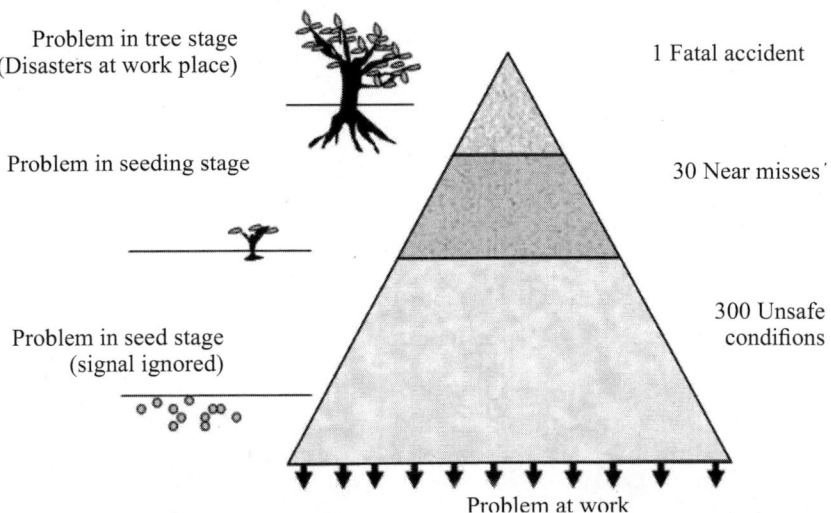

Figure 2.2 Theory of 1-30-300.

The unsafe condition in the beginning is easy to attend, such as a problem in the seed stage. But people ignore the signs while the unwanted seeds are spreading. The near misses are like a problem in the seedling stage. It is possible to remove the seedling with slight effort, but people do not find time to do that. Finally when it becomes a tree, it is not possible, or viable to remove. People learn to live with the problem or become a victim of it. Following are few examples of the unsafe conditions in textile mills:

By providing a belt guard, accidents can be prevented and workers would be safe. If a worker has removed the belt guard for some work, the supervisor should insist him to put it back. By not doing this, one might meet with an accident and loose his limbs or even his life.

Running the machines without belt guard is one of the main reasons for accidents in textile industry.

Machine running without a belt guard

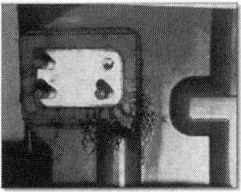

Loose and unsafe wires and electrical installation

Mills allowing the loose wiring, damaged insulations, etc., and concentrating only on getting production finally end up in a fire accidents.

By not storing the yarns properly in bags or cartons and not keeping them in a designated place, mill will not be able to get the required yarn as per count, shade, lot, etc., leading to production loss and quality problem. This problem can be solved by keeping the materials in designated place while receiving it.

Haphazard way of keeping material

By not covering the cones while the civil work is going on, the cones get damaged and it is not possible to salvage good materials. Covering the materials is the easiest solution to avoid this loss.

Civil work going on while materials are not covered

The remnant cones from warp creel should be collected shade wise and lot wise, and there shall be no problem of again sitting and segregating cones and chances of mix up.

Remnants not collected shade wise and lot wise

By not covering the bleached yarns and allowing them to become dirty results in higher yarn wastes and rejection of fabrics due to stains. Putting each cone in a polythene bag is easy and the same bags can be reused.

Improper handling of bleached yarns

The fire buckets are supposed to be kept clean and always filled with water so that during the fires they can be used. Even though they are small, they can extinguish fire and prevent major losses. Neglegancy by supervisors and their heads of the departments is the reason for such lapses.

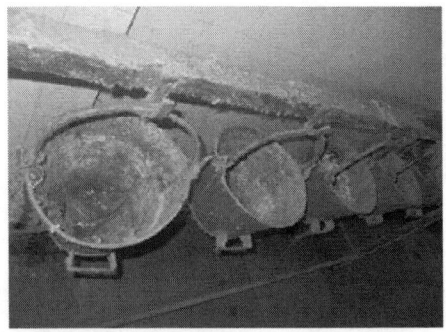

Empty fire buckets

Loading more material than the height of a person in a hand trolley pushed by a worker can result in accidents, as the worker pushing the trolley cannot see the road.

Loading more material than height of a man

Although a barrier is put in front of an electrical control panel, yet people are keeping materials inside. This can lead to major accidents. The supervisor in the shop floor is supposed to ensure that no materials are kept inside the barrier.

Keeping materials in front of electrical control panel

Figure 2.3 Unsafe conditions found in textile mills.

2.3   Find the root

For finding the root cause, we need to go deep in investigation by asking "Why" a number of times. This technique is known as "Five Why Technique" or "Why-Why-Why Technique." We need to first analyze the boundary of a problem. The next step is to do brainstorming with the concerned people and listing all the possible reasons for the problem. Then a cause and effect analysis is to be done to know which of the reasons are prevalent. Analysis of all the reasons by careful observation is made. Take the example of a breakdown of a shaft in a machine:

Q. Why this breakdown happened?

   Ans. The load on the shaft was very high.

Q. Why the load became high?

   Ans. The bearing was found jammed.

Q. Why the bearing got jammed?

   Ans. It was found dry without any grease.

Q. Why there was no grease?

   Ans. The grease nipple was found chocked?

Q. Why the grease nipple was chocked?

Ans. The mechanic did not clean the grease nipple during routine cleaning.

Q. Why he did not clean?

   Ans. Maintenance supervisor did not insist.

Ans. I had no system of verifying the maintenance work on regular basis.

Q. Then who is responsible?

Ans. Myself.

The same investigation may throw light on a number of shortcomings such as:

- The grease tins were not properly closed and dust and water entered in them
- Air had entered the grease gun and therefore, grease was not coming out when pumped
- Cotton fluff had entered the bearing area
- The machine was running at a higher speed than specified
- The maintenance schedule was not followed and the machine had worked continuously for a long time
- The fitter had given the work of greasing to a newly joined cleaner who was wandering outside
- Open bearing was used instead of sealed bearing to reduce cost
- Used grease was mixed with new grease to reduce cost
- A cheaper grease was used to save the cost

2.4 Preliminary questions

Preliminary questions is a technique that is essentially a development of Five Ws and a H, a checklist that is recommended for selective use. This is very helpful not only for root-cause analysis but also for developing procedures, systems, and new products.

Who

- Who is affected by the problem?
- Who else has it?
- Who says it is a problem?
- Who identified the problem first?
- Who says it is not a problem?
- Who would like a solution?
- Who would not like a solution?

- Who could prevent a solution?
- Who could add to the problem?
- Who need it solved more than you?
- Who handles this?
- Who is accountable?
- Who is worried about this?

When

- When it was observed first?
- When does it occur?
- When it does not occur?
- When did it appear?
- When will it disappear?
- When do other people see your problem as a problem?
- When others do not see your problem as a problem?
- When is the solution needed?
- When might it occur again?
- When will it get worse?
- When will it get better?
- When to take action?

Where

- Where it was noticed first?
- Where it is most noticeable?
- Where it is least noticeable?
- Where else does it exist?
- Which is the best place to begin looking for solutions?
- Where does it fit in the larger scheme of things?

Why

- Why is this situation a problem?
- Why do you want to solve it?
- Why others do not consider it as a problem?
- Why don't you want to solve it?

- Why doesn't it go away?
- Why would someone else want to solve it?
- Why wouldn't someone else want to solve it?
- Why is it easy to solve?
- Why is it hard to solve?

What

What might change it?

What are its main weaknesses?

What do you like about it?

What do you dislike about it?

What can be changed about it?

What can't be changed?

What do you know about it?

What you don't know about it?

What will it be like if it is solved?

What will it be like if it isn't solved?

What have you done in the past with similar problems?

What principles underlie it?

What values underlie it?

What problem elements are related to one another?

What assumptions are you making about it?

What seems to be most important about it?

What seems to be least important about it?

What are the sub-problems?

What are your major objectives in solving it?

What else do you need to know?

How

- How to start?
- How you would like to move?
- How to avoid this?
- How to convince others?

An example table

	Expected Gains				Expected Losses			
	For you		For others		For you		For others	
	Tangible	Subjective	Tangible	Subjective	Tangible	Subjective	Tangible	Subjective
Option 1								
Option 2								

- How to move further?
- How to arrange resources for solving this?
- How to project the problem to top management?
- How to end?

2.5 Personal balance sheet

A tree has a number of roots of different sizes and different penetration in the soil. Similarly, there may be a number of roots for a problem with different depth at different regions. It need not be necessary to remove all the roots, as few of them might not be effective when few other roots are removed. The personal balance sheet technique suggested by Janis and Mann can be used for this purpose. This technique was originally used by counselors, etc., for explaining people as to why a specific change was proposed. It was later developed as a part of problem solving technique. A decisional balance sheet or decision balance sheet is a tabular method for representing the pros and cons of different choices, and for helping someone decide what to do in a certain circumstances. They are often used in working with ambivalence in people who are engaged in behaviors that are harmful to their health.

The above technique becomes useful for finding the roots of a problem and also to take help from all, as we need cooperation from all involved and make them feel that the existing problem is harmful to them also. The typical examples of this are implementation of incentive schemes, modernization of plant and machinery, implementation of quality management system, re location of factory, and so on.

2.6 Plusses, potentials, and concerns

Firestien developed the concept of plusses, potentials, and concerns in 1989, a technique that constructively evaluates an idea and is closely related to the developmental response, receptivity to ideas, and advantages, limitations, and unique qualities. The development of each idea is quite time consuming and therefore, the technique is more appropriate for use on a short-list of ideas than for general screening of large number of ideas. This can help in self assessment like SWOT analysis.

The first step is to state the ideas into a simple form. Then following listing is done:

(1) Three or more 'plusses' (strong points)

(2) Three or more 'potentials' (spin-offs, researchable possibilities, etc.)

(3) Your 'concerns' about the idea, and putting them in order of importance

(4) Starting with the most important idea and making notes on how to overcome each concern (or at least the main ones)

(5) Taking into account step 4, trying to improve your original idea for instance:

    (a) How to get people to understand it and become enthusiastic for it

    (b) Its advantages and disadvantages (and how to surmount the disadvantages)

    (c) The resources required (people, materials, money, etc.)

    (d) How to pre-test it (e.g. are there particular times or locations you might use?)

    (e) How to identify when implementation is complete

(6) In order to keep the momentum going, put in place the opening steps of a suitable action plan, with at least one step to be done within the next day.

2.7 Problem reversal

This technique is explained in the book "What a Great Idea" by Charles Thompson. The theory behind this is, "the world is full of opposites." Of course, any attribute, concept, or idea is meaningless without its opposite. Lao-tzu wrote "Tao-Te-Ching" which stresses the need for the successful leader to see opposites all around. He stated that "the wise leader knows how to be creative." In order to lead, the leader learns to follow. In order to prosper, the leader learns to live simply. In both cases, it is the interaction that is creative. All behavior consists of opposites. Learn to see things backwards, inside out, and upside down.

The Method consists of the following steps:

(1) Make the statement negative: For example, if you are having a complaint of slubs in your yarn, work out how you can produce that slub intentionally on the machines you have without incorporating any new attachments. You may try with adding few long fibers, damaging few cots or aprons, reducing pressure on drafting, taking the top rollers front or back, and so on. If the defect produced intentionally resembles the complaint you received, you are clear about the root.

(2) Doing what everybody else doesn't: For example, when tight pants are prevalent in the market, introduce bell bottoms with a premium price;

when all are producing pants, introduce new type of printed dhotis as the main dress of a super star in a movie.

(3) The "what-if compass": The author (Charles Thompson) has a list of pairs of opposing actions that can be applied to the problem. Just ask yourself "What if I ..." and plug in each one of the opposites.

For example:

Stretch it/Shrink It

Freeze it/Melt it

Personalize it/De-personalize it, etc.

(4) Change the direction or location of your perspective: Physical change of perspective; manage by walking around, or doing something different.

(5) Flip-flop results: If you want to increase sales, think about decreasing them. What would you have to do?

(6) Turn defeat into victory or victory into defeat: If something turns out bad, think about the positive aspects of the situation. If I lost all of the files of this computer, what good would come out of it? Maybe I would spend more time with my family?! Who knows!

In textile and garment industry, the biggest problem faced is that the end customers always need something new. No one wants to wear a similar dress as others are wearing, unless it is a compulsion as uniform. They want to project themselves as different. The fashion designers and textile technologists are always working to produce something new. In this process, they get something that was not there in their agenda. Few call it a defect, and few others call it an effect. Majority of the fancy effect and fashion have come from the so called mistakes or defects. If one knows how to produce it in the same way as it is, then it becomes an effect. If he does not know how to reproduce that effect, then it is a defect. The classical examples are slub yarns, faded jeans, torn jeans, mélange yarns, crackers, corkscrew yarns, neppy yarns, loop yarns, and so on. In garment industry, any new fashion is like that.

After receiving a complaint if we can reproduce it somehow, we would be able to know the root cause. If that effect is not required, it can be removed. Reproducing a defect is not an easy job. A defect comes at random when everything is working right. Somewhere something goes wrong for a short duration of time, and getting a combination of all is very difficult.

2.8 Cut the roots

Whatever might be the method adopted to identify the root, cutting is more important without which the problem cannot be solved. If you leave the root, the problem is going to grow again. There are numerous methods of attacking the problem and solving it that are explained in the forthcoming chapters.

In textile and garment industry, majority of the problems are repetitive in nature indicating that the root cause analysis is not done and roots are removed fully. Even after knowing something is bad or the root of a problem, the same thing is insisted by the management and finally the problem is made to hit back. The classical examples are appointing people for the job without verifying their competency levels, booking the orders without verifying the capacity, assuming customer requirements without interacting with customers, telling lies to impress a customer, cooking figures to impress someone, attitude of neither learning nor educating the subordinates, harassing the subordinates to show one's supremacy, not studying the procedures and manuals while doing the works, not analyzing the complaints for finding root cause, not analyzing the non-conformities raised by auditor before taking action, taking decisions based on personal experience without referring to facts, not paying the people what they deserve, giving bribes to officials, and so on.

So many seven steps

3.1 Problem solving techniques

अग्निः शेषं ऋणः शेषं शत्रुः शेषं तथैव च
पुनः पुनः प्रवर्धेत तस्मात् शेषं न कारयेत्

Sanskrit Subhashita

*"Agnih shesham runah shesham shatruh shesham tathaiva cha punah
punah pravardheta tasmaat shesham na kaarayet"*

Fire, loan, and enemy if remain even in small traces will grow again and again, so finish them completely.

Problem like fire, loan, and enemy if left without tackling and allowing even a trace to remain will resurface again and hence, it is necessary to remove the problem by its root.

"Problem Solving" is a tool, a skill, and a process. It is a tool because it helps in solving an immediate problem or to achieve a goal. It is a skill that can be used repeatedly, such as adding numbers or speaking a language. It is a process that involves a number of steps. One can engage in problem solving to reach a goal or when experiencing obstacles on the way. The goals are likely to be many and varied. It is likely that in working towards his goals, one will encounter few barriers. On the point at which one comes up against a barrier, he can engage in a problem solving process to help achieve the goal. Every time a problem-solving process is used, the skill increases.

Effective problem solving is one of those skills that just cannot be forced. The harder you try the poorer your problem solving skills will become. The more stressed you become and the more time pressure you feel, the worse you will become at solving problems creatively. You may follow the same steps, but a man with cool mind can see the root of the problem, whereas others cannot.

Problem solving is a process to move from the existing situation towards the goal. It is a sequence of steps or actions taken to solve a problem including, defining the problem, identifying or creating possible solutions, and choosing among the solutions. Decision making is complementary to problem solving and refers to the act of selecting one or more options from those available; it does not involve creating possible options. Decision making may take place by default; that is, without consciously recognizing that an opportunity for decision making exists. This fact will lead us to the very first element in a definition for decision making. To have a decision making situation, there must be at least two alternatives. If there is only one course of action there will be no decision making, for there is nothing to decide.

The essence of continual improvement is the ability to solve problems effectively, but making it happen in a systematic and effective manner is much more difficult. Problem-solving skills rarely come naturally; they must be refined and practiced daily. But these skills can be developed by most individuals and organizations. There are so many techniques practiced world over; interestingly, a number of them are termed as "Seven Steps for Problem Solving." Few of the techniques are explained here.

3.1.1 Q.C. Story – Seven Steps for Problem Solving – Japanese system

There are a number of techniques for problem solving. The seven steps for problem solving, popularly known as QC Story, designed by Dr. Juran and Prof. Ishikawa, are relevant for all situations. They are as follows:

(1) *Problem identification* – Define the problem clearly

(2) *Observation* – Recognition of the features of the problem

(3) *Analysis* – Finding out the main causes

(4) *Action* – Action to eliminate the causes

(5) *Check* – Confirmation of the effectiveness of the action

(6) *Standardization* – Permanent elimination of the causes

(7) *Conclusion* – Review of the activities and planning for future work

Step 1: The first step, problem identification, consists of the following sub-steps:

(1) Highlighting the importance of the problem and the need to solve it

(2) Understanding the background of the problem and the course of actions taken so far

(3) Working out the present loss because of the problem

(4) Deciding on the extent to which the problem is required to be solved during the present task

(5) Sanctioning a budget for solving this problem

Use of as much data as possible is essential to identify the most important problem.

When a problem is selected, one must be sure of the reasons for this selection. The circumstances in which the problem gets priority are to be identified and highlighted. The undesirable results of the poor performance are to be expressed in concrete terms; especially in monitory terms in order to attract the attention of top management. If the degree of importance is extremely high and is widely understood by many people, the problem will be dealt seriously. Next, the loss in performance in the present situation and the advantage of effecting improvements are described. The basis on which the target values are set in the theme and how it is important are indicated. When the theme includes many kinds of problems, then they are to be divided into sub themes for effective handling of the problem. Brainstorming is often used to identify all possible reasons for the problem to occur.

For example, if low productivity in ring frames is the problem, an analysis shall be done broadly to identify the areas, as to whether it is poor working, shortage of back materials, shortage of empty bobbins, lack of training to workers, improper balancing of machines, poor condition of the ring frames, and poor quality of back materials, and so on. When we analyze, normally all reasons shall be there, but their contribution to the loss shall not be uniform. We need to work out as to which reason is responsible for most of the losses and attack it first.

Step 2: The second step, the observation, consists of investigation of the specific features of the problem from a wide range of different viewpoints. Four different viewpoints considered are time, place, type, and symptom. As the intensity and effect of problem depends on various factors, the investigation must be done from different points of view to discover variation in results. One needs to go to site and collect necessary information that is not put in the data form. The objective of this step is to discover the factors that are responsible for causing the problem.

For example, when observations were made for poor productivity in ring frames, the stoppage of machinery waiting for doff was found more in the shift end, the working problem was found more during interval hours, shortage of back process were found in the beginning of shift, shortage of skilled labor found in the evening shift, and so on.

In a number of cases it is found that the normal reason that is told by all shall not at all be there, but a new reason is identified which was never expected. A classic example is given below.

In a composite textile mill, there was a problem of holes in the fabric that was noticed after processing. There were no holes in grey fabric. The holes found were random in nature. They were found in different lots and with different process configurations like only bleached, dyed, printed, and coated, and so on. The processing people were blaming weaving and weavers were blaming processing. When the processing and weaving people were unable to find the root, the spinning master decided that he shall investigate and did not inform his plan of investigation. He came in the night at around 11.30 PM and went through the complete process from grey inspection.

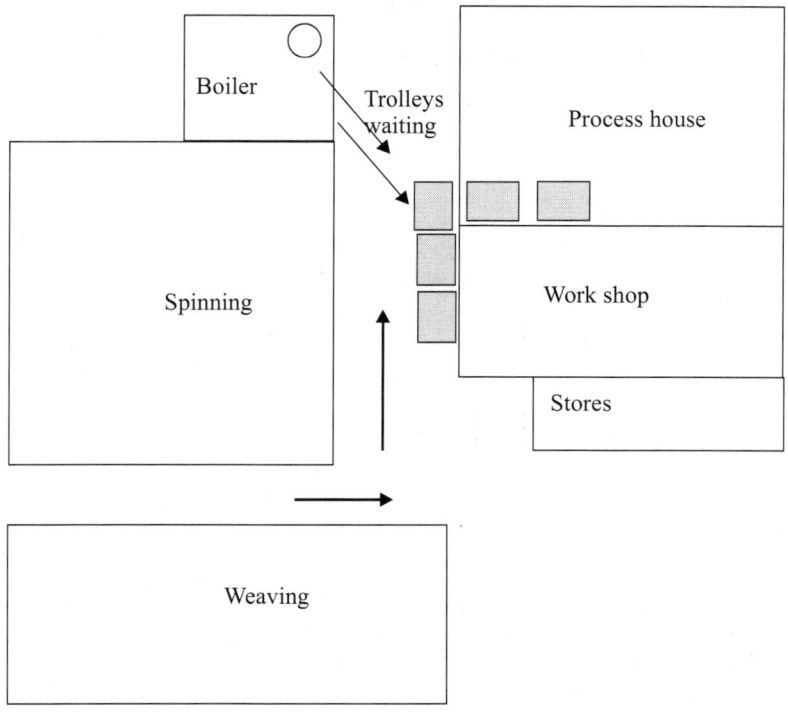

Figure 3.1 Floor plan.

The weaving and processing were located away from one another by 200 m as shown in the Figure 3.1. In the end of second shift at 10.00 PM, all the fabric trolleys were brought from weaving and kept in queue near the process house as shown. The spinning master observed burning particles coming out from the boiler chimney and falling on the road where trolleys were kept in queue. After noticing the problem, immediate arrangements were made to cover the cloth trolleys with tarpaulins.

Step 3: The third step is to find out the main cause of the problem. The activities include setting up of hypothesis, verification of all links using cause and effect diagram, deleting the elements that are not relevant and testing the hypothesis to identify the real root of the problem. Use of cause and effect diagram is to ensure the collection of all possible knowledge concerning causes. The information obtained in the observation step is used to delete the elements which are not clearly relevant. Then the cause and effect diagram is revised by marking the elements that have a high possibility of being the main causes. New plans are devised and data is collected to ascertain the effect of those elements on the problem. The data is validated by small experiments, and sometimes poor result is intentionally reproduced.

The problem of holes in processed fabrics was found due to burning particles falling from boiler chimney on the grey fabric placed outside the process house. This problem was not there earlier as the only coal was being used in the boiler, but recently the mills had started using sugarcane bagasse wastes procured from a nearby sugar factory, as the coal was in shortage.

The hypothesis set before analyzing the problem was:

(a) There may be some rough surface on the metal emery used in looms

(b) Some chemical brought in powder form is not dissolving fully

(c) Something wrong might be happening in stenter

(d) Too much stretch given in stenters, and so on

However after the observation all hypotheses were changed, as the problem was due to burning particles coming out of the chimney.

Step 4: The fourth step, action, is to eliminate the main causes. A strict distinction must be made between actions taken to cure phenomena (immediate remedy) and actions taken to eliminate casual factors (preventing recurrence). The ideal way of solving a problem is to prevent it from happening again by adopting remedies to eliminate the cause of the problem. While taking action, care should be taken to ensure that the actions do not produce other problems. If it is inevitable, devise remedies for the side effects. The action has to be thoroughly evaluated and judged from a wide range of view points as possible. It is advisable to conduct trial and check method. Devise number of different proposals for action; examine the advantages and disadvantages of each, and select those which the people involved agree to. An important practical point in selecting actions is ensuring the active cooperation of all those involved.

The actions that were possible to prevent the problem of holes in the fabric are to cover the trollies, to stop using bagasse in boiler, to cover the complete roads, and so on. Among them, the management found that covering the trollies was simple, effective, and less expensive. The action could be taken immediately as tarpaulins were there in the mills brought to cover cotton bales, trucks, etc.

Step 5: Step five is the checking to ensure that the problem is prevented from occurring again. This involves comparing the results before and after the implementation of the solution. The data should be collected in the same format as it was done for analyzing the problem, and comparison is to be made for before and after. It is also essential to relate the situations before and after for understanding. Then effects are converted into monitory terms and compared with target value. If the undesirable results continue to occur even after actions have been taken, the problem solving has failed. We need to go back to the observation step and start again.

Step 6: The sixth step, standardization, is to eliminate the cause of the problem permanently by devising the procedures to perform the activities and documenting them. Without documenting the procedures and standardization, the actions taken to solve the problem will gradually revert to the old ways and lead to recurrence of the problem again, and also it is likely to revert when new people are appointed on work.

This is a normal problem found in the majority of textile mills because of which, even after running the mills for a long time, a number of problems are still there and are getting repeated. Every time a new man comes, he would like to do the same trials again that were done earlier and proved ineffective.

Standardization will not be achieved simply by documents. It must become a part of the thoughts and habits of the management, staff and workers. Education and training to all involved along with assigning responsibilities is an important part of this step.

Step 7: The last step is reviewing the problem solving procedure and planning future work. The activities involved are summing up the problem remaining, planning for solving the remaining part of problem, and verifying what went well and not well. One should realize that a problem is never perfectly solved and an ideal situation almost never exists. It is not good to aim for perfection or to continue the same activities on the same theme for too long.

It might not be possible to have the same system and parameters all over the mills because of the differences in the type of machines available, the types of raw materials being used, the count combination, and so on. One has to identify the similarities in system and implement the corrective actions suitably. For example, increasing relative humidity might had given good result while spinning short cottons for a coarse count, but might result in lapping in superfine cottons with high honeydew contents. A large shuttle may be helpful in coarse counts as it can hold more yarns, but in fine counts, it may result in higher warp breaks. A large sized Simplex bobbin may be helpful in coarse counts as it reduces a number of creel changes, whereas the same shall not be suitable for superfine counts.

When the original time limit is reached, delimiting the activities is important. Even if the target is not reached, a list should be made of how far the activities have progressed and what has not been attained yet. It might be worthwhile to live with a problem rather than eradicating it fully; which depends on the nature of the problem remaining, its after effects, the steps and cost needed to eradicate it fully.

A mill may decide to provide residential quarters to their maintenance fitters in order to get their services fast in case of breakdowns, whereas they may find it not viable to provide quarters for all although it can help in reducing absenteeism and employee attrition.

A systematic approach as mentioned in seven steps above reduces the number of problems and helps in moving towards zero.

3.1.2   Seven-step model from Mycoted

This is a modified version of PDCA adopted for problem solving. Four steps are in Plan, one each in Do, Check, and Act. The steps are as follows:

Plan

(1) Identification of the problem or the target to be achieved which can add value (pulse value).There is no meaning in simply taking a problem, by solving which no value is added.

(2) Understanding the nature of the problem or the activities and the risks involved. Whether the problem is technology related, human practice related, human psychology related, human value related, raw material related, time related, etc., need to be understood before taking a problem for solving.

(3) Identifying and verifying the root causes. Constructing a cause and effect diagram, reviewing and identifying the root cause. Studying whether a riskfree method or a process can be achieved.

(4) To take decision as per situation; often referred as 'new action' or 'date value'. Develop a solution and an action plan. Generate potential solutions, rank these and then generate the tasks to deliver the solution. Construct a detailed plan.

Do

(1) Implementation of the solution. Communicating the plan and reviewing the plan regularly amongst all concerned. Involving the people concerned is very important part of this.

Check

 (1) Reviewing and evaluating. Using the performance measures identified to review and evaluating the results of the change. One need to ensure whether the result obtained was really because of the action taken, or some other factor is playing in between.

Act

 (1) Reflecting and acting on what was learnt. Assessing the problem-solving process to obtain lessons learnt. Continuing the improvement process where needed.

3.1.3 Creativity, innovation, and problem solving

This method explained by Peter Sylvan in "Quantum Books" consists of understanding the customer requirements in detail, including the problems being faced and the actions taken, the success or failures, asking right questions and making a proper problem statement, developing proper tools and procedures, getting good ideas from everyone, serendipity (process for coming up with useful new ideas), searching for multiple solutions, referring to patents and project notes, etc.

Understanding the customer problems from the point of view of customer is the most important step; and hence, working with the customer in his work place is suggested. Use of fictitious product descriptions to stimulate ideas and discussion are also adopted when required. Asking the question repeatedly with different words and viewpoints, trying graphical terms, describing the problem to a layman and also to experts in different fields and verifying the reactions, studying inverse problems or transforming one problem into another, and trying making it worse are few of the tools suggested for understanding the problem in depth.

Creative problem solving depends on using the right tools, tricks, procedures, or methods of analysis. In few cases, new tools and methods of analysis must be developed from scratch by the inventor before a problem can be solved; and in other cases, special tools and procedures must be developed to take the final critical step of enabling successful commercial applications.

Serendipity, an aptitude for making desirable discoveries by accident, is a very effective process for coming up with useful new ideas, but requires one to keep the eyes open and imagination turned on like learning from Mother Nature (the originator of serendipity) and studying the lessons or investigating any unexplained phenomena she may reveal. Useful solutions are sought by reviewing backlog of problems while browsing at random in libraries, trade shows, and the real world. Reviewing problems before going to sleep at night and keeping a notepad and audio recorder handy are helpful. Meditation under a tree or in an open field, playing with combinations of ideas and concepts, thinking about analogies to the problem are also used in this technique.

Searching for multiple solutions is recommended as the first solution is found usually inadequate or non-optimum, and sometimes having not more than one solution is considered dangerous. Studying multiple problems jointly often generate unique solutions. It is suggested that even if you have one optimum solution, it may be necessary to get patent coverage for all other effective solutions so as to protect your market. Value of experimentation, play, exaggeration, and persistence are given importance in this technique. It is suggested to spend some time in trying things which we normally do not do, or we do not know how to do. If you don't fail frequently, you aren't trying hard enough and may be missing a lot of good opportunities. It is suggested to be very stubborn about solving a problem, but be flexible about the definition of the true problem and open minded about the form of the solution.

Patent notebooks are used to provide legal protection for inventions, but can have many other useful, complementary functions such as a recorder, a reminder, a source of ideas, a means of ensuring project continuity, and a way to communicate within self and within a project group. Neatness is not essential, but clarity and conformance to legal standards is critical. Other things that should be recorded are sources, questions, what doesn't work, and the things to try. A truly effective and comprehensive patent requires planning, team work, and iteration. You can invite everyone to participate in finding ways around your patent claims or to break them or improve on them. A one page summary sheet of the important procedures and checkpoints should be included inside the front cover of every patent notebook issued.

When we consider an example in textiles of a white bleached fabric, color contamination is not accepted. But the same contamination if made to repeat randomly and more frequently, the fabric fetches a premium price. If the contamination is multi-colored, the value added is much higher. What you need to do? Collect hard wastes of different shades, open them separately and contaminate them after carding in the sliver.

3.1.4 A seven-step problem solving cycle – University of South Australia

A seven-step model given by the University of South Australia that mainly targets students explains that there are a variety of problem-solving processes, but each process consists of a series of steps including identifying an issue, searching for options, and putting a possible solution into action. It is useful to view problem solving as a cycle because, sometimes, a problem needs several attempts to solve it, or the problem changes. Figure 3.2 shows a seven-step problem solving cycle.

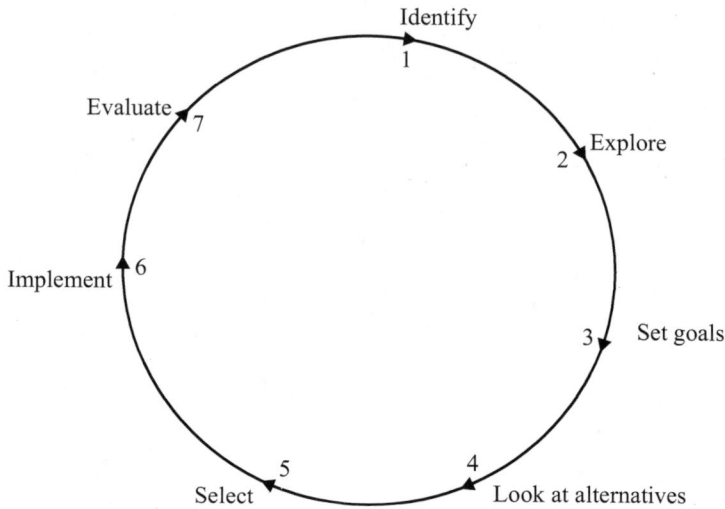

Figure 3.2 Seven-step model by University of South Australia.

Step 1: Identify the problem: The first step is to identify and name the problem, so that an appropriate solution could be found. Sometimes one might be unsure about what the problem is: might just feel general anxiety or be confused about what is getting in the way of goals. For personal problem one can ask self, friends, or a counselor.

Step 2: Explore the problem: When one is clear about what the problem is, the next step is to think about it in different ways. One can ask questions such as "How is this problem affecting me?", "How is it affecting others?", "Who else experiences this problem?", and "What did they do about it?" Seeing the problem in different ways is likely to help finding an effective solution.

Step 3: Set goals: Once the problem is seen from different angles, one can identify the goals. What is expected to achieve? Sometimes one might get so frustrated by a problem that he forgets about what he wanted at the first place. For example, he might become ill, struggle to complete a number of assignments on time, and feel so unmotivated that let due dates pass. It is important at this time to consider the question, what is the immediate goal: whether to improve health or time management skills, or complete the assignments to the best of the ability, or finish the assignments as soon as possible? Depending on the goals, the solution can be decided. So working out goals is a vital part of the problem-solving process.

Step 4: Look at alternatives: Once the goal is decided, we need to look for possible solutions. More possible solutions mean that we are more likely or able to discover an effective solution. You can *brain-storm* for ideas. Few of the best solutions arise from creative thinking during brain-storming. One can also seek ideas about possible solutions from friends, family, a partner, a counselor, a lecturer, books, or the internet. The aim is to collect as many alternative solutions as possible.

Step 5: Select a possible solution: From the list of possible solutions, one can sort out which are most relevant to the situation and which are realistic and manageable. One can do this by predicting outcomes for possible solutions and also checking with other people what they think outcomes might be.

Step 6: Implement a possible solution: Once a possible solution is selected, it is ready to put into action. One might need to have energy and motivation to do this because implementing the solution may take some time and effort, and prepare self to implement the solution by planning when and how to do it, whether to talk with others about it, and what rewards to be given when it is done.

Step 7: Evaluate: Just because one has worked through the problem-solving process, it does not mean that automatically the problem is solved. Evaluating the effectiveness of solution is very important. One can ask, "How effective was that solution?", "Did it achieve what I wanted?", and "What consequences did it have on my situation?" If the solution was successful in helping solving problem and reaching the goal, then one can conclude that the problem is solved effectively. If one feels dissatisfied with the result, then he can begin the steps again for alternative possibilities by beginning the problem solving cycle again.

3.1.5 Seven steps by Harry Joiner

Harry Joiner, an executive recruiter, gives seven steps for problem solving. When solving problems whether in real life or in a job interview, it is important to follow a logical process. Therefore, the strength of a job applicant's problem solving ability can be seen by walking them through the following seven-step framework while getting them to describe how they solved a real life problem in their last job. When discussing a problem that they solved in a previous job, the applicant should demonstrate the ability to perform the following:

(1) Defining the problem

(2) Defining the objectives

(3) Generating alternatives

(4) Developing an action plan

(5) Troubleshooting

(6) Communicating

(7) Implementing

Drilling down on how a candidate has solved problems in the past will give a good idea of how they will solve problems in the future in terms of the quality, consistency, and costs of their solutions. It is therefore suggested to verify this point while recruiting people for key positions.

3.1.6 Pushing through the problem – By small group communications

Seven steps of problem solving incorporating six steps for decision making is a unique method suggested by small group communications. The steps for problem solving are as follows:

(1) Defining and identifying the problem

(2) Analyzing the problem

(3) Identifying possible solutions

(4) Selecting the best solutions

(5) Evaluating solutions

(6) Developing an action plan

(7) Implementing the solution

The idea generated techniques such as brainstorming, buzz groups, nominal groups, Delphi methods, fantasy chaining, focus groups, and metaphorical thinking are included in identifying the possible solutions.

Step 1: Define and identify the problem: Defining and identifying the problem is a critical step. It is essential for each group member to clearly understand the problem so that all the energy is focused in the same direction. It is suggested to define the problem by writing down a concise statement that summarizes the problem, and then to write down the goal. The objective is to get as much information as possible. It may be helpful to divide the symptoms into hard and soft data. Hard data includes facts, statistics, goals, time factors, history, whereas soft data includes feelings, opinions, human factors, attitudes, frustrations, personality conflicts, behaviors, hearsay, and intuition. Sometimes information needs to be gathered via various devices like interviews, statistics, questionnaires, technical experiments, check sheets, brainstorming, and focus groups to define the problem.

Developing a problem statement is essential, which clearly describes the current condition which the group wishes to change. The problem defined should be limited in scope so that it is small enough to realistically tackle and solve. Writing the statement will ensure that everyone can understand exactly what the problem is. It is important to avoid such problems including any "implied cause" or "implied solution" in the problem statement. There is a saying, "a problem well-stated is a problem half solved".

Once the problem is defined, it is relatively easy to decide what the goal will be. Stating the goal provides a focus and direction for the group. A measurable goal will allow the tracking of progress as the problem is solved.

When defining the problem, ask the following:

- Is the problem stated objectively using only the facts?
- Is the scope of the problem limited enough for the group to handle?
- Will all who read it understand the same meaning of the problem?
- Does the statement include "implied causes" or "implied solutions?
- Has the "desired state" been described in measurable terms?

Do you have a target date identified?

Step 2: Analyze the problem: In this stage of problem solving, questions should be asked, information gathered, filtered, and sorted. We should not make any assumptions. The problem is to be viewed from a variety of viewpoints and not just how it affects us but also to others. It is essential to spend some time researching the problem either by going to the library or developing a survey to gather the necessary information.

The following questions are suggested to ask when analyzing the problem:

- What is the history of the problem? How long has it existed?
- How serious is the problem?
- What are the causes of the problem?
- What are the effects of the problem?
- What are the symptoms of the problem?
- What methods does the group already have for dealing with the problem?
- What are the limitations of those methods?
- How much freedom does the group have in gathering information and attempting to solve the problem?
- What obstacles keep the group from achieving the goal?
- Can the problem be divided into sub problems for definition and analysis?

Step 3: Identifying possible solutions: Using a variety of creative techniques, group participants create an extensive list of possible solutions. Asking each group member for input ensures that all viewpoints will be considered. When the group agrees that every course of action on the list will be considered, they will feel some direct ownership in the decision making process. This may help put the group in the mood of generating consensus later in the decision making process. This step is also referred as idea generation or finding optional solution.

Step 4: Selecting the best solution: For a problem, there might be a number of possible solutions and all can give results, but few are costly, few are irritating, few are not liked by many, few are simple, few are foolproof, few have some loopholes, and so on. One needs to discuss with the people implementing the solution and select the one convenient to majority.

Step 5: Evaluating solutions: Few look simple, just like "belling a cat" or "milking a tiger" but may not be as simple as we think. So identify the steps involved, processes involved, hurdles expected, the critical path, and evaluate the solutions. In textiles, balancing the machines and working full without loss of utilization and efficiency while having the flexibility of changing to the needs of fashion is one such thing. People talk of monthly projections, annual projections, and even five year plans, whereas the looms shall be idle for want of programme. Balancing the product mix, balancing the market segments, targeting people with huge money, vacating an established market, increasing market share by taking over other units are few of the solutions given for improving the financial position of a mill; but out of these suggestions, what can be implemented successfully is the question.

Step 6: Developing an action plan: Giving suggestion is very easy as compared to implementing it. One needs to prepare an action plan considering all hurdles and supporting elements in order to achieve the result by implementing the plan. Consider various tools like PERT, PDPC, FMEA, QFD, and develop an action plan that is robust, and implementable.

Step 7: Implementing the solution: The results can be obtained only when a concept is implemented fully. After preparing action plan, one needs to educate the people involved, guide them the way you want it to be implemented, provide required resource and system, supervise the implementation, and ensure consistency in implementation. In order to get the result, involvement of people is very important, or else the implementation shall not be successful.

Verify whether the problem was solved or still existing. It is normally found that a problem is partially solved and there shall still be a lot to attend.

3.1.7 Gravel Gulch – Four steps to problem solving

Dr. Win Wenger gives a four-step model instead of popular seven steps. It is named as Gravel Gulch. The basic procedure is to break the problem-solving process into a series of discrete and specific steps, then move systematically through each step in turn. Each of these steps, in turn, involves a "brainstorm" to look at as many possibilities and aspects as possible; then reduce and narrow all that down, selecting from among these the basis for the next step. The four steps are as follows:

Step 1: The "mess", also known as the "blow-off," a cathartic elaboration of all that can be said about the problem situation. Out of that mess, select or create the one statement which best defines the problem or question. This statement might be quite different from your initial statement of what the problem is. Make as many roughly one-sentence statements as you can about the problem "mess" or situation. These can be factual statements, as implied by the "fact-finding" name. These can also be feelings you have about the issue, and everything else that's wrong or bothersome in that context. In fact, quite literally, anything you can think to say about this mess or problem situation, say it! If it occurs to you in the context, write it down whether you think it fits or not. Write all the things you think, feel, perceive and know about this situation. Write as rapidly as possible as, faster than judgment can possibly hope to keep up with, perhaps 50-100 entries in 10 to 15 minutes. You need to outrun your internal editor so that all kinds of remarkable new insights and perspectives can open up.

Step 2: Idea-finding is done by brainstorming dozens or even hundreds of possible remedies to that problem statement, from which 1 or 2 most interesting possibilities are selected. In a fixed time of 10–15 minutes, list 30–50 possible entries, many silly, many that at first look silly but on closer inspection may be invaluable instead. Write whatever comes into your mind, even if it seems to have nothing to do with the solution.

In a number of cases, it is seen that the points considered as silly or seemed as nothing to do with the solution becomes the real idea by which that problem can be solved. For example, extinguishing a fire in oil well by blasting, putting bucket full of water in coal-fired boiler when fire is not enough, burning the fields to prevent forest fire from entering the fields, and so on.

Step 3: Solution-finding is brainstorming dozens or scores of ways to turn that interesting idea into an actual solution to the problem from which you select the one or two ways that most nearly appear to actually lead toward the desired results. Expand your best idea or combination of ideas into three to four sentences description (or prescription), showing how that idea or suggested course of action will solve the problem.

Step 4: Action-planning, which, depending upon context, may involve any or several discrete steps in planning, in finding acceptance, in implementation.

3.1.8 Seven steps for problem solving by Tim Hicks

Tim Hicks, founder and director of Connexus Conflict Management, gives following seven steps for problem solving. He says that there are two important things to remember about problems and conflicts: they happen all the time and they are opportunities to improve the system and the relationships.

Here are seven steps for an effective problem-solving process:

(1) Identify the issues.

- Be clear about what the problem is.
- Remember that different people might have different views of what the issues are.
- Separate the listing of issues from the identification of interests.

(2) Understand everyone's interests.

- This is a critical step that's usually missing.
- Interests are the needs that you want satisfied by any given solution.
- The best solution is the one that satisfies everyone's interests.
- This is the time for active listening.
- Separate the naming of interests from the listing of solutions.

(3) List the possible options.

- This is the time to do some brainstorming. There may be lots of room for creativity.
- Separate the listing of options from the evaluation of the options.

(4) Evaluate the options.

- What are the pluses and minuses?
- Separate the evaluation of options from the selection of options.

(5) Select an option or options.

- What's the best option, in the balance?
- Is there a way to "bundle" a number of options for a more satisfactory solution?

(6) Document the agreement(s).

- Don't rely on memory.
- Writing it down will help you think through all the details and implications.

(7) Agree on contingencies, monitoring, and evaluation.

- Conditions may change. Make contingency agreements.
- How will you monitor compliance?
- Create opportunities to evaluate the agreements and their implementation.

Effective problem solving takes time and attention, but less than a problem not well-solved. Also solving a problem might cost something, but much less than living with the problem.

3.1.9 Kepner and Tregoe method – Seven steps each for problem analysis and decision making

This technique emphasizes the "rational" rather than the "creative", it is essentially a method for fault diagnosis and repair rather than for disorganized or systemic problem domains, or those where freshness of vision is essential. Kepner and Tregoe (1981) describe the method below, but its origins date from the 1950's.

The method is fully developed, with recommended techniques, worksheets, training programmes, etc. The headings below provide a bare outline and follow two main stages, each has seven steps:_

Problem Analysis

(1) You should know what was ought to be happening and what is happening. This can then be expressed as a deviation; compare them and recognize a difference that seems important to you.

(2) Ascertain provisional problem priorities (how urgent/serious or likely to become so) and pick a problem to work on. Break down unhelpful problem categories. If the cause is immediately apparent, you can pass straight to decision making.

(3) Investigate and identify the problem deviation (what, where, when, and to what extent).

(4) Identify features that distinguish what the problem is from what it is not.

(5) Identify the potential cause(s) or contributory factors of the problem, these should be clear-cut events or changes that lead to

the problem and are clearly associated with the occurrence of the problem.

(6) Attempt to infer any likely causes of the problem by developing hypotheses that would explain how the potential cause(s) could have caused the observed problem.

(7) Now test the potential cause of the problem by checking that it is not only a potential cause but also that it is the only cause (e.g., that occurrence of this problem is always and only associated with occurrence of this cause or combination of causes).

Decision making

(1) Set up specific requirements:

- Expected results (what type, how much, where, when)

- Resource constraints (personnel, money, materials, time, power, etc.)

(2) Prioritise your needs (distinguishing "musts" and "wants")

(3) Develop optional supplies of action, i.e. systematically investigating each requirement and identifying ways of accomplishing it. Alternatively, other idea generation methods could be used.

(4) Rate the alternatives against requirement priorities.

(5) Choose the best option as a provisional solution.

(6) Identify potential unfavorable consequences.

(7) Plan implementation including minimizing adverse consequences and monitoring progress.

3.1.10 PIPS – The Phases of Integrated Problem Solving

PIPS technique was developed by Morris and Sashkin in 1978. Its phases are a variation of the classic Creative Problem Solving (CPS) method. Here in addition to defining the range of analytic steps required, it also defines the inter-personal actions needed for each step, as shown in the table below:

	Problem-solving tasks	Inter-personal tasks
1. Problem definition	Search for information about the problem. Detailed understanding of problem situation. Agree on group goals.	Does the information search involve everyone? Open sharing of problem information. Consensus building.

	Problem-solving tasks	Inter-personal tasks
2. Solution generation	Brainstorm ideas. Elaborate and refine ideas. Develop tentative list of solutions.	Encourage all to brainstorm. Encourage no criticism. Encourage co-operation when listing solutions.
3. Ideas into action	Evaluate strengths/weakness of each idea. Try combining good ideas. Select a tentative solution.	Avoid non-productive criticism. Resolve conflicts over combining/ modifying ideas. Consensus building.
4. Action planning	List steps needed for implementation. Identify resources needed. Assign responsibilities for each step.	All participate in listing steps. Group adequately evaluates available resources. Develop real commitments.
5. Plan evaluation	Success measures for each step. Timetable to measure progress. Contingency planning in case steps need modifying.	All contribute to developing success measures. All comfortable with time-table. Real commitments for contingency plans.
6. Evaluate product and process	How well do effects of solution match original goals? Identify any new problems created. Any future actions needed?	How much group participation overall? Are self-expression and offers of support easy? What has group learned about itself?

To work effectively, PIPS requires a problem-solving group, a Facilitator, an observer to monitor the problem-solving tasks, and an observer to monitor the inter-personal tasks.

In theory, the observer's roles should be rotated, in as much as, at the end of each phase the previous observers would swap with others in the problem-solving group. The authors of the PIPS technique also provide a more detailed questionnaire than the table above, that all participants have for reference. The observers fill in the questionnaire. There is a general review at the closing stage of each step of the process issues, and members go on to the next step only when all the tasks of the previous step have been satisfactorily completed.

The complete PIPS process is almost certainly too cumbersome for routine problem solving, but may prove beneficial for training.

3.1.11 Problem solving stages by Kent Academy

Kent Academy explains that there are several stages in problem solving that are explained in simple terms as follows.

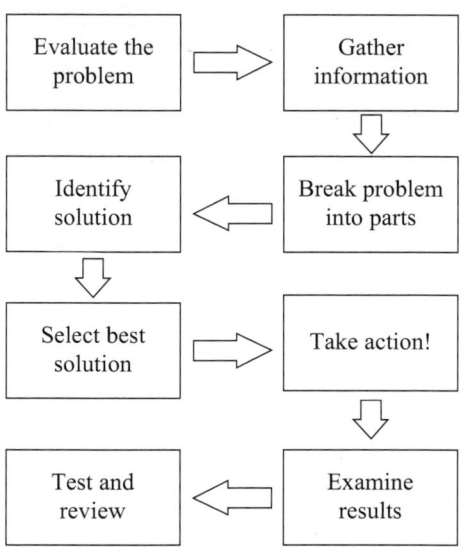

Figure 3.3 Different stages of problem solving.

(1) Evaluating the problem

   • *Clarifying* the nature of a problem

   • *Formulating* questions

   • *Gathering* information systematically

   • *Collating* and organizing data

   • *Condensing* and summarizing information

   • *Defining* the desired objective

(2) Managing the problem

   • *Using the information gathered* effectively

   • *Breaking down a problem* into smaller, more manageable parts

   • Using techniques such as *brainstorming* and lateral thinking to consider options

   • *Analyzing these options* in greater depth

   • *Identifying steps that can be taken* to achieve the objective

(3) Decision-making

- *Deciding between the possible options* for what action to take
- *Deciding on further information* to be gathered before taking action
- *Deciding on resources* (time, funding, staff, etc.) to be allocated to this problem

(4) Resolving the problem

- *Implementing action*
- *Providing information* to other stakeholders; delegating tasks
- *Reviewing progress*

(5) Examining the results

- *Monitoring the outcome* of the action taken
- *Reviewing the problem and problem-solving process* to avoid similar situations in future

At any stage of this process, it may be necessary to return to an earlier stage; for example, if further problems arise or if a solution does not appear to be working as desired.

3.1.12 Reductionism

Reductionism is a type of philosophy that can be applied to the problem solving process. It basically states that complex objects can be simplified in a way that makes them easier to understand. It is generally used to refer to scientific problems, but it can also be used for any problem that one encounters. The reductionism thought process can be defined in the expression, "complexity equals compound simplicity."

Many people find that they have a hard time solving problems because they try to tackle the problem head on. A much better method of solving problems is to reduce them to a fundamental level that makes them easier to solve. Once the problem has been broken down, one can look at the parts that make up the composition to understand the nature of the problem. Methodological reductionism states that the scientific ideas should be broken down into the smallest possible units but should not be broken down further than that. When we talk of finding the roots of a problem, it is actually applying reductionism and breaking the problem into small parts.

If we take an example of absenteeism in a textile mill, the reason may not be same in all the sections. The absenteeism in spinning may be due to poor working whereas in winding, it may be due to a bad supervisor. Again in spinning, the absenteeism might be high in certain group of people who all belong to a particular community migrated from a particular area. When the problems are broken and made into small pieces, action can be taken bit by bit and we start getting results. This concept is also a foundation for Kaizen where, small but consistent improvements are insisted rather than leaps and bounds.

3.1.13 Trial and error

Few complex problems can be solved by a technique that is called trial and error. Trial and error is typically good for problems where you have multiple chances to get the correct solution. However, this is not a good technique for problems that don't give you multiple chances to find a solution. In textile industry it is very common, especially with old timers, to try trial and error method for solving working problems. Where there are multiple variables contributing for a problem and a multiple combinations are available for solving the problem, this technique is used. For example, a variation in humidity and temperature, the variation in cotton, the settings in drafting, the twist parameters, the spindle speed, etc., all play a role in getting good working, whereas one cannot change all. One tries to do something and keeps up working good, may be by reducing the speed slightly or by increasing the twist. Sometimes some samples are to be produced and one cannot change all the configurations just for the sample, and tries some short cuts.

The trial and error is also a great way to gain knowledge. Basically, a person that uses the trial and error method will try this method to see if it is a good solution. If it is not a good solution, he would try another option. If the method works, the person using it would acquire the correct solution to the problem. However, there are few situations where there are too many options, and it is not feasible for a person to go through all of them to find out which one works the best. In this event, a person will want to use the option that has the best possible chance of success. If this doesn't work, they can try the next best option until they find a good solution.

There are a number of important factors that make trial and error a good tool to use for solving the problems. The purpose of trial and error is not to find why a problem was solved, but it is primarily used to solve the problem itself. While this may be good in few fields, it may not work so well in others. For example, while trial and error may be excellent for finding solutions to mechanical or engineering problems, it may not be good for certain fields which ask "why" a solution works. It is primarily good for fields where the solution is the most important factor. Hence in textile industry which is driven by fashion and managed by merchants, trial and error is a common technique adapted. Sometimes it is like gambling, and if it is favorable, results in a high profit.

A good aspect of the trial and error method is that it does not try to use a solution as a way of solving more than one problem. It is primarily used to find a single solution to a single problem. Trial and error is not a method of finding the best solution, nor is it a method of finding all the solutions. It is a problem solving technique that is simply used to find a solution. One of the most powerful advantages to this technique is that it does not require a person to have a lot of knowledge. However, it may require one to have large amounts of patience.

It is also an excellent tool for inventors. The inventors will first imagine a device they would like to invent, and then they may go through the trial and error process to find the best ways of inventing the device. While it is an extremely powerful tool that can be used to solve problems, it also has few weaknesses. It can lead to disasters if the combination selected is not favorable and hence in modern times, people are suggested to study the problem in detail, find the root cause, and then take the corrective steps.

3.2   Six problem-solving fundamentals

Craig Cochran explains six fundamentals of problem solving. They are as follows:

(1) Using a structured problem-solving method

(2) Assigning ownership of the problem

(3) Involving people familiar with the problem

(4) Applying project management techniques

(5) Aggressively pursuing the root cause

(6) Communicating, communicating, and communicating

Use of structured problem-solving method has the following advantages:

• Prevents problem solvers from jumping to conclusions.

• Ensures root cause analysis.

• Demystifies the problem-solving process.

• Prescribes which analytical tools to use and when.

3.2.1 Use a structured problem-solving method

There are many structured problem-solving methods along with the numerous analytical tools. Few are copyrighted, few are public domain, few are very intricate, and others are quite simple. Typically, they range in complexity from four to eight steps, but all the methods share the same basic themes. Therefore, it is less important which problem-solving method you choose than actually picking one and using it. One can even make up his own method. The structured seven steps suggested by Craig Cochran are as follows:

Step 1: Decide on which problem to pursue: It is a very important step. In most organizations there are countless opportunities for improvement but finite resources available to apply to them. The organizations must therefore prioritize the issues and dedicate resources accordingly. Appropriate tools for this step include brainstorming, Pareto charts, run charts, pie charts, flowcharts, and voting.

The management in any textile mill or garment factory would like to solve the problem that is costing more to the company. Therefore, develop a habit of working out the losses per year in monitory terms for each problem you have and prioritize on the maximum loss maker.

When mills are making profit, they say that there are no problems and start investing and expanding the activities, and fail because the recession follows the bloom and by the time the expansion is complete, the recession has started. Therefore even when you are making profit, identify the factors contributing for low profit, and work to correct the situation.

Step 2: Define the problem: In the clearest and briefest terms possible, explain what exactly the problem is. Provide the details of who, what, where, and when. Keep in mind that carefully defining the problem will provide the raw material for successfully identifying its root cause. Appropriate tools include brainstorming, Pareto charts, check sheets, and histograms.

When textile people are asked about the performance, they normally say "market is bad" and never say "I failed to analyze the trends and predict the market in spite of my experience in the field for so many years." They do not accept that they were not proactive to face a situation of recession, which is cyclic in nature. By hiding a problem and telling something else can never help in problem solving.

Step 3: Determine the root cause: Identifying a root cause is confused a number of times with defining the problem itself. This is because of mistaking a symptom for the root cause. Often the so-called "root cause" is nothing more than a restatement of the problem definition. Before team members are asked to participate in problem solving, they should receive training in how to distinguish symptoms from root causes (Appropriate tools include interviewing, brainstorming, cause-and-effect diagrams, and voting).

Let us make a simple exercise for root cause analysis:

Loom efficiency is low

- *Why?* – Weft breakages are high.
- *Why?* – When the cone size becomes small, it causes more breaks.
- *Why?* – The paper cone is getting compressed by the pressure of the yarn.
- *Why?* – The paper cones are reused to reduce the cost.
- *Why?* – The top management is not convinced to invest on plastic cones.
- *Why?* – The technical person is not able to convince the top management because of the figures of loss caused due to the poor efficiency of looms as compared to the savings done by using old paper cones.
- *Why?* – The technical person has not yet developed his ability to communicate and present the actual situation to the management.
- *Why?* – The technical person was not educated and trained for managerial skills.
- *Why?* – The technical person expected the management to sponsor for his training, and never made his own effort to learn.
- *Why?* – The technical person felt it as the responsibility of management to train him as the company was getting benefit by his training, whereas management was not ready to spend money for training.
- *Why?* – Management insists that they are recruiting people with high caliber and paying high salaries to them; and hence, it is the responsibility of the candidates to learn and implement the new systems. If management has to spend for training the senior persons, then what is the use of their qualification. Also, management argues that the employees after getting training start blackmailing them by saying that they have got a better offer and also insist for increments and promotions.
- *Why?* – A number of companies adopt a shortcut of snatching competent trained persons from different companies, rather than training their own employees to reduce the cost.
- *Why?* – The people at the top are not convinced about training and bringing their people up, as the people are not loyal and leave the company as soon as they get a higher salary offer.

- *Why?* – The people do not trust each other.

- *Why?* – Someone cheated the management because of his greediness earlier and hence, the management does not trust anyone.

- *Why?* – The management did not fulfill the requirements and did not recognize the people contributing and hence, people have no trust in the management.

The root cause for low efficiency at loom was not really the old paper cones used, but the lack of knowledge, lack of trust, and lack of team work. The decision made for using old paper cones was not with a bad intention, but a result of lack of knowledge. Similarly in that mill, there might be hundreds of such wrong decisions. Developing team work, encouraging people to work, and developing trust among the people can overcome all those problems.

Step 4: Generate possible solutions and choose the most likely one: This step works very well in a team setting to generate a large number of alternative solutions. The trick is to cast a wide net, then narrow the possibilities to those solutions that satisfy the following criteria:

- They have a strong chance of being successfully implemented.

- They will be accepted by all relevant stakeholders.

- They truly address the root cause identified in the previous step.

Then agree upon a solution, either by group consensus or through executive decree (appropriate tools include brainstorming, Pareto charts, and voting).

Let us consider the example of technical staff leaving the company. To curb this, a number of solutions can be identified by having a brain storm. Following are few of the solutions listed after a brainstorming session:

(1) Always keep your salary level higher as compared to the salaries given in other mills for the same level and work load.

(2) Take a bond and security deposit from the employees for minimum working period of ten years; if they leave within that period, they will lose the deposit and the interest.

(3) Go on conducting staff development programme, so that the people stay with you to learn.

(4) Provide good living quarters with facilities, so that the wives shall insist their husbands not to leave the quarters.

(5) Do not recruit any one for higher post; let recruitments be only for lower posts and people get promotions within the company.

(6) Keep people fixed at small areas and provide higher salary so that they become unacceptable in other mills.

(7) Do not conduct any training programme and do not provide time for people to think or read, so that they become unacceptable to others and stay with you only.

(8) Close all communication systems so that your people cannot get information on the developments taking place elsewhere and therefore, they cannot think of joining other companies.

(9) Do not send your staff to any conference or seminar where they have an opportunity to meet people from other mills.

(10) Discourage people from becoming members of professional bodies like The Textile Association (India), as they come in contact with people working in other mills.

(11) Recruit only dull people who do the work as told to them without using their brains.

(12) Do not procure any newspaper or magazines for the company as people may see the advertisements and job opportunities.

(13) Make an agreement with fellow mill owners that they will not drag people from your mill and you will not drag people from other mills.

(14) Employ only illiterate or less educated people for higher and technical jobs as they cannot dare leaving the company.

(15) Increase the notice period in case of resignations and retrenchments from the present 3 months to 6 months or one year.

(16) Make a policy of offering jobs to the kith and kins of employees only when they have served for over five years in your company.

(17) Offer job opportunities to the children of employees who have served for over 20 years in your company.

(18) Provide schooling facility to the children of your employees at a good reputed school for which the fee shall be borne by the company.

(19) Insist employees to bring along their families and provide jobs to their spouses.

(20) Give annual maintenance contract to outsiders so that your staff remains incompetent and not attracted by other companies.

(21) Encourage marriages among the children/brothers/sisters of your staff within the circle of the staff.

(22) Encourage your relatives to join your company for the technical posts.

(23) Always go on installing latest technology so that technicians are attracted and they do not leave your company.

(24) Install transparent management systems so that people shall be happy to work with you.

(25) Treat your staff as your own family members and develop love and affection with them.

(26) Make a policy of providing jobs to the dependents in case an employee dies or becomes physically challenged due to any reason.

(27) Make a policy of not retrenching any staff member for poor performance as you have selected them as the best among the people appeared for interview, and it is your responsibility to make them competent if they are not.

(28) No boss can harass his staff; he has to work with them to get the desired results.

(29) All selections must be done by a team consisting of senior functional heads, and no single person should be allowed to recommend any person from among his relatives.

(30) No head of the department should have the power to accept a resignation. All resignations should be forwarded to the managing director or chief executive who discusses with the resigning person, finds out the reason for his resignation, and accepts it only when he is convinced that the person resigning is doing so only because of some other reason which is beyond the control of the management.

(31) Arrange social gatherings among the staff and their families and build better relationship between bosses and subordinates.

(32) Provide special increments depending on the number of years of service.

(33) Provide loans to the staff for constructing houses.

(34) Provide different loans like car loan, loan for purchasing furniture, education loan for higher studies, loan for home appliances, etc., and see that the loans are deducted in installments.

(35) Encourage staff to become shareholders of the company by offering shares at face value or with slight premium, but cheaper than market rate.

(36) Encourage employees to start cooperative society and take active part in it.

(37) Encourage employees to form a cooperative housing society and construct apartments for them to live.

(38) Develop a reserve force; nothing will happen if someone leaves.

(39) Treat people well and they will not leave you.

(40) Show your plans for future and make people a part of it.

(41) Stick to the principles and people shall stick to you.

(42)Do not break your own rules.

Choosing the most appropriate solution from among various solutions suggested in brainstorming is a real tough job as all are having few plus points and few minus points; although, all have the same purpose of retaining the staff. Few need investment, whereas few need determination and change of attitude. Few suggestions are calling for a reverse thinking against the basic purpose of the business.

Step 5: Plan and execute the solution: Even the best solution is doomed to fail if its implementation isn't carefully planned and executed. This process typically consists of two distinct phases: selling the solution to key stakeholders in order to get buy-in, and methodical project planning to ensure that the solution is executed correctly. It is also helpful to notify the organization's customers who will be affected by the solution. This reinforces the idea that the organization is dedicated to customer satisfaction and problem resolution (appropriate tools include project planning, effective presentation skills, selling skills and pilot runs).

One of the main reasons for the failure in implementation of good solution is because of non-cooperation by fellow members; although they are the one to get the maximum benefit. The main reason for non-cooperation is the fear that the credit of implementation might go to the initiator and to them. By involving the people and presenting them as core committee members or advisors for the project, the work can be implemented; even though these people do not contribute anything for the implementation, yet they will not disturb the process. In Sanskrit, there is a *subhashita* as follows:

दुर्जनम् प्रथमम् वन्दे सज्जनम् तदनन्तरम्
गुद प्रक्षोपलम् प्रथमम् मुखम् तदनन्तरम् च

[*durjanam prathamam vande sajjanam tadanantaram*
guda prakshopalam prathamam mukham tadanantaram cha]

It means, "First wish the crooked; good people can be wished later. The ass hole gets washed first and then the face."

Step 6: Verify effectiveness: After implementation of the solution, someone must verify that it is effective. It isn't absolutely necessary that people outside the problem-solving team verify its effectiveness, but it might be helpful in order to avoid bias. Whether they are internal or external, customers are particularly good at shedding light in this regard. If a customer doesn't perceive an improvement, then there is no improvement. Perception is everything (appropriate tools include auditing, interviewing, documentation, control charts and process capability).

It is always advisable to involve the customer in verifying the results and their validation. It not only helps in getting the correct solution but also curtails customers from making fake allegations.

In textiles we normally see that the customers during the recession period send complaints while saying that the quality is bad in spite of the material being made exactly as per their specified requirements, whereas they accept everything during boom period, although the quality is bad. This attitude brings pressure on technical staff and they are unnecessarily harassed by marketing personnel and also the top management while making the lives of technicians miserable, and forcing them to search for an alternate job for none of their mistake.

Step 7: Communicate and congratulate: This step is routinely forgotten in many organizations. People crave information about how problems are being addressed and solved. This information creates a feeling of security and confidence, and builds a culture of continual improvement. Recognition is also important. People who successfully contribute to problem-solving efforts should be recognized for their work. Congratulations should be dignified i.e., public and carried out by top management (appropriate tools include empathy, integrity, and effective speaking and writing skills).

In a number of cases in textile mills it is seen that the one who is really instrumental for an improvement is not recognized, but the head of the department without contributing a single percent comes forward to take the reward on behalf of the department. The head of the department does not even spell out the names of the people who worked for improvement, whereas it is necessary to identify the person who gave the idea or concept, and the people who cooperated with him in making the concept workable.

Although an organization uses a team approach for problem solving, yet every problem should be assigned to a specific individual and should be confirmed that this person accepts the ownership. The owner is simply the project manager for solving that particular problem. Being selected as "problem owner" in no way indicates accusation or blame but in fact, it is a vote of confidence in the person's ability as a leader and manager. Individuals who are accountable for projects lead the vast majority of successful problem-solving projects. Ownership can make remarkable things happen; it should not be neglected.

3.2.2 Involve people familiar with the problem

Those most familiar persons with the variables surrounding a problem should be involved in the problem-solving process. They need not be managers and supervisors but may be anyone like people taking orders, writing software, operating machines, driving forklifts, and performing repairs. An organization's culture must allow all personnel to contribute actively to the process, regardless of their level within the organization.

One of the project manager's most important tasks will be to select the right people for the problem-solving team. Participants should be told why they've been included (e.g., their technical expertise, familiarity with processes in question, or experience in the department). It is important that the individuals are motivated and enthusiastic about being involved.

One of the problems observed in textile mills is that the staff assumes that workers are having lesser knowledge as compared to them. They do not have any patience to listen to the workers. Of course, a worker may not be able to explain his problem in simple terms or in the terms the manager can understand, but it is the responsibility of the manager to step down one or two steps and try to understand what the worker is trying to tell. Once you listen to a worker, the worker gets himself involved in the project you are taking forward.

3.2.3 Apply project management techniques

Project management is a very basic concept consisting of assigning responsibilities, timeframes, milestones, and reviews, and then tracking them to completion. A well-designed, corrective, and preventive action system embodies the basics of project management. The system should be user-friendly and streamlined; then it's perfectly suited as a project-management tool for problem solving.

3.2.4 Aggressively pursue the root cause

An explicit step of almost all problem-solving models is the identification of the root cause. But just because it is explicit, it doesn't mean it will happen. Identifying a problem's true root cause must be encouraged, and it's the project manager's responsibility to see that this is done. It is not easy, and it usually takes some serious investigation and intellectual tenacity. Keep in mind that a root cause is rarely the first thing that comes to mind.

3.2.5 Communicate, communicate, and communicate

It is suggested to make problem-solving success stories a frequent subject within the organization. If a customer complaint gets addressed effectively, it should be published in the company newsletter. If a group of employees succeed in reducing the error rate, it is advisable to send everyone an e-mail trumpeting the achievement. If the quality assurance department assists a supplier in improving the consistency of the output, publish it in the local newspaper to cover the story. Get the word out any time your organization succeeds in solving or preventing problems. The more often employees hear about successes, the more they will want to be involved. And the more they become involved, the more successful the company will become.

Dignified public recognition is, of course, a form of communication, one that delivers an astronomical return on investment. The message underlying public recognition is, "the company appreciates your team's fine efforts, and we sincerely hope others will follow your example." Who wouldn't want to follow their example and be recognized too?

Communicate, communicate, and communicate? In a number of mills, it is seen that communication is not effective. What is the meaning of it? There are slogans written all around, the procedures are documented, messages are displayed, the training classes are conducted, email messages are sent, the matter is explained not only on phone but also personally, a highly paid consultant explains the same thing a number of times, even then people do not change their old methods, do not maintain the records as suggested by the management representative, or the quality control incharge. They say that they were not aware of the system. But when the boss insists and takes action on people not following the system, they start implementing. The communication means not only telling them but making them understand that if they do not follow the system, they will not be spared.

3.2.6 Conclusion

There should be an end to all problems. Do not go on dragging the same problem all the time, otherwise you will not get the support from others. People will get fed up with you, and they would consider you as a problem and work for eliminating you instead.

3.3 Beyond methods – Twenty points to help you solve problems

Dr. Win Wenger developed a checklist to aid problems in everything from day-to-day life to the most abstract of situations. They are as follows:

(1) You need to be firm, and ensure that you really want to solve the problem. In a number of cases, it is seen that the one pointing the problem is not keen to solve it but want to use it as an excuse to cover his inefficiencies. For example, shortage of skilled workers, higher absenteeism, lack of knowledge among the staff, recession in the market, and fluctuations in raw material prices, and are few such problems. If you are really interested in solving these problems, highlight your efforts and the results you got rather than blaming some invisible person for this, and use it as an excuse for the poor efficiency, poor quality, unorganized house, improper record keeping, not taking actions as decided in a meeting, and so on.

(2) Have wide-ranging interests and feed them. If you concentrate only on certain group, others will try to run away from you.

(3) Be willing to entertain ideas and inspirations from inside the box and not only "think outside the box." Learn from any and every source as per our new ancient saying: *"Anyone can learn from someone wise; it takes someone pretty wise to be able to learn even from fools."*

There is a saying in Sanskrit:

अमन्त्रम् अक्षरम् नास्ति मूलम् अनौषदम्

आयोग्यम् पुरूषम् नास्तिए योजकस्तद दुर्लभः

Amantram aksharam naasti, naasti moolam anoushadam,

Ayogyam purusham naasti, yojakastada durlabhah

(1) There is not a syllable which is not a spell; there is not a plant that is not a medicine; there is no person who is useless; only, the harnesser is hard to find.

(2) It means, on this earth there is no letter without meaning, no plant without medicinal properties, no man who is useless, but we have the scarcity of the people who can plan and use them for the best. Each one has his own strengths and one should know the skill of using them. Similarly, each one has few weaknesses also. There are opportunities for all and also the threats. One needs to analyze and understand them.

(3) In textile industry, majority of failures are due to not identifying the value of people working and not organizing them properly to get the results.

(4) Be willing to keep coming back to the problem from different directions. By seeing the problem from only one direction, you might not be able to understand it. People see problems in different ways and

have their own perceptions. You need to understand them all so as to get cooperation from all.

(5) Be willing to let go of it between times, deal with other matters like tending the garden, washing the dishes, meditation, experiencing, or doing some art work, or taking inordinate pleasure in little things.

(6) Keep/build your stamina and follow-through. Do not lose interest in the task as it makes you physically weak. If you are physically weak, you cannot achieve success.

(7) Maintain your health.

(8) Keep your day job.

(9) Improve your sense of humor. It reduces tension and you would be able to think.

(10) Be fully creative, *then* fully critical, *then* fully creative…

(11) Raise and keep up your level of ongoing tinkering:

    (a) Tinker with the problem.

    (b) Tinker with the idea or with ideas.

    (c) Tinker with other things.

    (d) Be opportunistic.

    (e) Fiddle in other creative activities and keep those further resources of yours in the picture.

(12) Work in creative bursts; don't make it routine office/factory work. Grinding a chapter a day just doesn't do it. Fly on inspiration as fast as possible before the pattern dissipates. Fly fast on inspiration as long as possible, then climb right back and go up again. You get more of what you reinforce. Moreover, the unique rewards of working will inspire and keep you reinforced to be creative. Be willing to dog-plod few of the tasks on some sort of scheduled regular basis of production, but do as much as possible by being inspired. Don't wait for inspiration to come, find it.

(13) Build high self-esteem:

    (a) Reinforce your confidence by being self-critical from time to time. If you are convinced on what you are thinking and doing, you can convince others.

    (b) Search hard for everything that might be wrong with your idea-theory-discovery-invention, then: "Damn the torpedoes, full steam ahead!" Do not assume that whatever you think is always correct. Think again and again and you will find lots of flaws.

(14) Do your homework; keep on getting better informed in the context.

(15) Pat yourself on the back at few of those many occasions when no one is going to do that for you. Find others also doing something worthwhile and pat them on the back. A small but definite percentage will reciprocate. Find/create a support group. You don't have to be alone. Support can be found in unexpected places.

(16) Appreciate "the closest distance between two points" in human affairs is usually a very zigzag line!

(17) Appreciate the assets and abilities which have brought you this far already.

(18) Appreciate the many, many others who have been part of this road and somehow made it through. Resolve to be with them and not with those who instead fell to the wayside. You deserve to make it through, you're going to make it through, you have it within you and above you what it takes to make it through! People need, human beings depend on, what you're bringing through! And many with far less potential than you have made it through!

(19) Be sure of at least some of the worth of what you are seeking to bring through.

(20) Get visibly on record everyone who says "no" to you and their grounds for saying "no." Prepare for publication of your running memoirs about your campaign and how these people, by name, title, and position, said "no." Few will find it safer to say "yes" rather than join the public ranks of the following, as reported once in Reader's Digest under "History Lessons."

Let us move further to understand various methods of identifying the roots of the problems and getting ideas to overcome them.

Storm your brain

4.1 What is brainstorming?

Brainstorming is a group or individual creativity technique by which efforts are made to find a conclusion for a specific problem by gathering a list of ideas spontaneously contributed by its member(s). The term was popularized by Alex Faickney Osborn in the 1953 book "*Applied Imagination.*" Osborn claimed that brainstorming was more effective in a group, rather than individuals working alone for generating ideas; although more recent research has questioned this conclusion. Today, the term is used as a catch for all group ideation sessions.

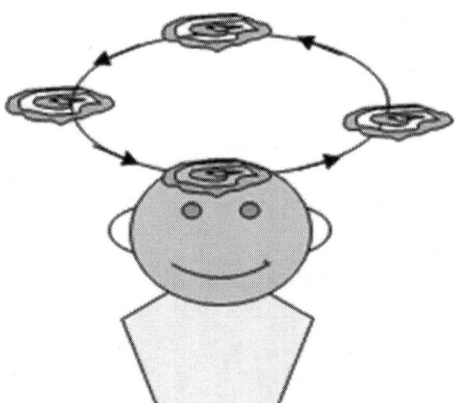

Figure 4.1 Storming the brain.

The brain is the organ that can think and help in solving all the problems. If one consciously takes advantage of his natural thinking processes by gathering brain's energies into a "storm," he can transform these energies into written words or diagrams that will lead to lively and vibrant writing. The concentration on a subject while digging out all the possible and relevant facts and imaginations can lead to the discovery of a number of hidden facts.

Brainstorming stirs up the dust, whips some air into stilled pools of thought, and gets the breeze of inspiration flowing again. Brainstorming techniques can range from picking words out of the dictionary randomly to considering things from another person's point of view, or even filling out a specifically designed chart. It is particularly helpful when one needs to break out of stale established patterns of thinking, so as to develop new ways of looking at things.

Brainstorming is used widely to help a group in creating as many ideas as possible in a short time, while trying to identify the possible causes of a problem, possible effects of a decision taken, and alternate methods for implementing a system. Whether one is starting with too much information or not enough, brainstorming can help to put a new task in motion or revive a project that hasn't reached completion.

Brainstorming combines a relaxed, informal approach to problem solving with lateral thinking. It encourages people to come up with thoughts and ideas that can, at first, seem a bit crazy. Few of these ideas can be crafted into original and creative solutions to a problem, while others can spark even more ideas. This helps to get people unstuck by "jolting" them out of their normal ways of thinking.

Brainstorming works by focusing on a problem, and then deliberately coming up with as many solutions as possible, and by pushing the ideas as far as possible. One of the reasons it is so effective is that the brainstormers not only come up with new ideas in a session, but also spark off from associations with other people's ideas by developing and refining them.

Brainstorming, or free associating, is usually considered to be thinking of random ideas to work towards generating a solution to a problem. This is also used for developing new opportunities in order to improve the service offered. This increases the richness of solutions explored. There are times when there is too much chaos in the brain and there is a need to bring in some conscious order. In such cases, brainstorming forces the mental chaos and random thoughts to rain out onto the page, giving some concrete words or schemas that can then be arranged according to their logical relations.

Brain when not utilized is like a pond of stagnant water that looks very clean from the surface. If a churning is made, lot of things that were at the bottom shall surface out. Few may be useful and few may not be. One can collect the useful items and progress from them. Similarly when brainstorming is done, it is actually churning of the stagnant brain. Lots of thoughts come out. Few are relevant and few irrelevant. Collect the relevant information.

Brainstorming is normally a group creativity technique designed to generate a large number of ideas for the solution to a problem, whereas it can be done individually also.

Brainstorming is one of the most creative ways of solving problems. However, brainstorming may not always be used to solve a single problem. It is mostly used to come up with a list of possible solutions that can be used to solve a problem.

4.2 Brainstorming history

The exact beginnings of brainstorming are not recorded, probably because brainstorming is a creative thought process that comes to people naturally. The mention of brainstorming is there in Indian mythology. The demon *Mahishaasura* (महिषासुर) was very powerful, and to get rid of him the divine people had a *"Vichara Manthana"* (विचार मंथन) [*Vichara* = thoughts, *Manthana* (Also referred as *Mathana*) = churning] and worked out a strategy to kill him. Goddess *Chaamundeswari* (चामुण्डेश्वरि) was created and each one gave his/her powers to her, so that she could win and bring peace. If this was an illustration of group brainstorming, the case of *Narasimha* is that of individual brainstorming. The demon *Hiranyakashipu* (हिरण्यकशिपु) had a number of favors granted to him by Lord *Brahma*. He was to be killed neither by man nor by animals, neither during day nor in night, neither inside the house nor outside the house, and neither by weapons nor by fire, and so on. So the Lord *Narasimha*, who was neither a man nor an animal, a human body with the head of Lion, came out from a pillar, just when the sun was setting (neither day nor night), caught the demon, and killed him at the doorstep, neither inside the house nor outside. He used his claws, and not any weapons or fire. We do not know whether all these events happened or not, but there is a clear message of generating creative ideas using brainstorming. These stories are compiled in *Puraanaas* (पुराण) by Maharishi *Vedavyasa* (वेदव्यास) who was in *Dwaaparayuga* (द्वापरयुग) in about 4000BC, and these stories are said to have taken place about 10,000 years ago.

While the fundamentals of brainstorming have been in use throughout the history, Alex Faickney Osborn, a 1940s advertising executive and one of the founders of BBDO, developed guidelines to conduct the brainstorming, and the modern brainstorming technique was developed. He found that the conventional method of overcoming obstacles and creating new ideas were too inhibitive and were not conducive to real creativity. The method was first popularized in a book called *"Applied Imagination."* Osborn proposed that groups could double their creative output by using the method of brainstorming. He referred to brainstorming as a "think up" process that had four fundamental rules, which stated that:

- The goal of a "think up" session would be to come up with as many ideas as possible.

- There would be absolutely no criticism of any thoughts or ideas.

- No idea should be considered too outlandish, and such ideas would be encouraged.

- Members of a "think up" team should build upon one another's ideas.

After Osborn introduced the concept of brainstorming, it took the world by fire. Prof. Juran with the help of Prof. Ishikawa and Dr. Deming made this technique simple and popular, which was used extensively in the Quality Circle movements, and Kaizen initiatives at Japan. This created a revolution. Nowadays, companies across the globe benefit from brainstorming and use it for problem solving, improvement initiatives, marketing concepts, advertising campaigns, and management methods and strategies, and so on for many more purposes.

The word brainstorming normally refers to a group activity, although individual brainstorming is also in practice. The sages, who developed the science, normally thought alone in a remote areas, which was normally referred as *Dhyana* (ध्यान) or *Tapas* (तपस) in Sanskrit. They used to stay away from human dwellings and concentrated on one subject on a continual basis. The birth of classical epics like Ramayana, Mahabharata, and a number of others is the result of concentrated efforts of churning the thoughts from the brain. The great finding of Lord Budha, "The root cause for grief is the desire" was due to his *tapas*.

While working to find a solution to a problem, normally one man cannot have the complete experience or knowledge of the situation and so it is necessary to involve all concerned from various sections where the roots of the problem are spread. The subject is to be made clear and specific to the participants, as it helps to focus their thoughts and ideas. Each member tells one reason at a time in rotation, and if he/she is not ready, shall say "pass", and allow the next person to speak. There shall be no discussions while the points are being put up, and no one will laugh or comment on the points put up by others. There is no need of giving any explanation or justification while the points are being told. All points are just recorded on a black board or a flip chart to avoid repetition. This is a very good group-education technique, eliminates bias to some extent, and brings a feeling of oneness in the group or team as the participants sit together while sharing their experiences through ideas. Brainstorming is a lateral thinking process. It asks that people come up with ideas and thoughts that seem at first to be a bit shocking or crazy. One can then change and improve them into ideas that are useful, and often stunningly original.

Although brainstorming has become a popular group technique, researchers have generally failed to find evidence of its effectiveness for enhancing either quantity or quality of ideas generated, especially when the subject discussed is technical or scientific in nature, which demands deep knowledge of the subject. Because of problems such as distraction, social loafing, evaluation apprehension, and production blocking, brainstorming groups are little more effective than other types of groups, and they are actually less effective than individuals working independently. For this reason, there have been numerous attempts to improve brainstorming, or to replace it with more effective variations of the basic technique. Although traditional brainstorming may or may not increase the productivity of groups, it has other potential benefits, such as enhancing the enjoyment of group work and improving morale. It may also serve as a useful exercise for team building.

4.3   Individual brainstorming

When one brainstorms on his own, that too in calm atmosphere, he will tend to produce a wider range of ideas than with group brainstorming as he does not have to worry about other people's egos or opinions, and can therefore be more freely creative. However, one without much experience or basic knowledge might not develop ideas as effectively as a group with experience. Individual brainstorming puts one in complete control of the creative process, and that means he is solely responsible for any and all results. This makes individual brainstorming a welcome challenge for few, and a source of concern for others.

"Individual brainstorming" is the use of brainstorming in solitary. It typically includes such techniques as free writing, free speaking, word association, and drawing a mind map, which is a visual note-taking technique in which people draw their thoughts. Individual brainstorming is a useful method in creative writing, and has been shown to be superior to traditional group brainstorming. Research has shown individual brainstorming to be more effective in idea-generation than group brainstorming.

Figure 4.2 Individual brainstorming.

One is most creative when he is relaxed, and most incredible creativity occurs when one is asleep i.e., when he dreams. At these times, brain creates three-dimensional, multi-sensory experience in real time. Imagine if one had to do that while he is awake! And this is why hypnosis is more effective for boosting creative problem solving skills. One can access that dream state to order!

4.3.1 The upside of individual brainstorming

The mindtools.com explains that while individual brainstorming may not allow one to take advantage of the accumulated experience of other members of the team, it will provide one with the freedom to express ideas without fear of ridicule or rejection. An idea that one might have been hesitant to bring up in a group brainstorming session may come to realization during the individual brainstorming process, and that single idea may be the one that makes the process a success. In addition to providing with added personal freedom, individual brainstorming also forces one to dig into the brainstorming process and give self over to it entirely. When an individual brainstorms with a group of people, he or she may be inclined to allow others to lead the process. When individual brainstorming is being conducted, there is no one else to rely on, which motivates the brainstormer to generate ideas and concepts on their own. While there is much to be said for group brainstorming, individual brainstorming is a process that should not be overlooked. There is much to be said for having the freedom to manage the creative process without the influence of opinions, ideas, or egos of others. While brainstorming on your own, it can be helpful to use mind maps to arrange and develop ideas.

It has been experienced that if group brain storming is to be more effective, the people should have individual brain storming before the session, and list all their ideas on a paper, which will help in putting up the points without delay.

4.4   Group brainstorming

Group brainstorming can be very effective as it uses the experience and creativity of all members of the group. When individual members reach their limit on an idea, another member's creativity and experience can take the idea to the next stage. Therefore, group brainstorming tends to develop ideas in more depth than individual brainstorming.

Brainstorming in a group can be risky for individuals. Valuable but strange suggestions may appear stupid at first sight. Because of such, a leader needs to chair sessions tightly so that uncreative people do not crush these ideas and leave group members feeling humiliated. While those who like to retain complete control of the creative process may find group brainstorming frustrating. Group brainstorming offers advantages that are not available when the individuals face the creative process on their own.

4.4.1 Benefits of group brainstorming

As a rule, people generally have creative boundaries that they stay within. Oftentimes, these boundaries are referred to as a "box," and when one begins to think "outside the box," the creative process can really take off. Group brainstorming helps the members of the brainstorming team to think outside of their boxes, and opening creative doors for each member of the brainstorming team.

When brainstorming is conducted alone, a person has only his sole knowledge and experience to rely upon. By creating a diverse brainstorming team, multiple founts of knowledge and experience will be available for the team to draw. Group brainstorming can also act as a team-building exercise by making sure that all the members of the team express their opinions and contribute their ideas. Group brainstorming is about creativity, productivity, and development. When there are a number of personalities and perspectives at work, the results of the brainstorming process can't be helped but be maximized.

4.4.2 How to use tool

To run a group brainstorming session effectively, following steps are to be followed:

(1) Define the problem to be solved clearly, and lay out any criteria to be met; If possible, give the problem in advance so that each one can have individual brainstorm before coming to the brainstorming session.

(2) Keep the session focused on the problem; do not allow the subject to deviate.

(3) Ensure that no one criticizes or evaluates ideas during the session. Criticism introduces an element of risk for group members when putting forward an idea. This stifles creativity and cripples the free-running nature of a good brainstorming session.

(4) Encourage an enthusiastic, uncritical attitude among members of the group. Try to get everyone to contribute and develop their ideas, including the quietest members of the group.

(5) Let people have fun while brainstorming. Encourage them to come up with as many ideas as possible, from solidly practical ones to wildly impractical ones. Welcome creativity.

(6) Ensure that no train of thought is followed for too long; if you find it happening, break it by interrupting, telling jokes or incidence, and divert the minds. Then again ask them to continue.

(7) Encourage people to develop other people's ideas, or to use other ideas to create new ones.

(8) Appoint one person to note down ideas which come out of the session. A good way of doing this is to use a flip chart. This should be studied and evaluated after the session.

Where possible, participants in the brainstorming process should come from as wide a range of disciplines as possible. This brings a broad range of experience to the session, and helps to make it more creative.

4.5　General outline for brainstorming

There are four basic rules in brainstorming. These are intended to reduce the social inhibitions that occur in groups and therefore, stimulate the generation of new ideas. The expected result is a dynamic synergy that will dramatically increase the creativity of the group.

(1) Focus on quantity: This rule is a means of enhancing divergent production, aiming to facilitate problem solving through the maxim, *quantity breeds quality*. The assumption is that the greater the number of ideas generated, the greater the chance of producing a radical and effective solution.

(2) No criticism: It is often emphasized that in group brainstorming, criticism should be put 'on hold'. Instead of immediately stating what might be wrong with an idea, the participants should focus on extending or adding to it, while reserving criticism for a later 'critical stage' of the process. By suspending judgment, one creates a supportive atmosphere where participants feel free to generate unusual ideas.

(3) Unusual ideas are welcome: To get a good and long list of ideas, unusual ideas are welcomed. They may open new ways of thinking and provide better solutions than regular ideas. They can be generated by looking from another perspective or setting aside the assumptions.

(4) Combine and improve ideas: Good ideas can be combined to form a single very good idea, as suggested by the slogan "1 + 1 = 3". This approach is assumed to lead to better and more complete ideas, rather

than merely generating new ideas alone. It is believed to stimulate the building of ideas by a process of association.

4.5.1 Set the problem

One of the most important things to do before a session is to define the problem. The problem must be clear, not too big, and captured in a definite question such as, *"What service for quality assurance is not available now, but needed? Or what are the possible reasons for a shade variation while dyeing on a HTHP dyeing machine?"* If the problem is too big, the leader should divide it into smaller components and each with its own question. For example, "what can make this shaft break while turning on a lathe? What can make this shaft break while working in a machine? What can make this shaft break while restarting a machine? Few problems are multi-dimensional and non-quantified; for example, *"what are the aspects involved in making a customer satisfied"*. Finding solutions for this kind of problem can be done with morphological analysis.

4.5.2 Create a background memo

The background memo is the invitation and informational letter for the participants, containing the session name, problem, time, date, and place. The problem is described in the form of a question, and few example ideas are given. The ideas are solutions to the problem, and used when the session slows down or goes off-track. The memo is sent to the participants at least two days in advance so that they can think about the problem beforehand and come prepared with their views. This shall reduce the wastage of time during the collection of the ideas.

4.5.3 Select participants

The leader composes the brainstorming panel consisting of the participants and an idea collector. Usually, a group leader is designated to write down the ideas. Ten or fewer group members are generally more productive than larger groups. Many variations are possible, but the following composition is suggested:

- Several core members of the project who have proved themselves
- Several guests from outside the project, with affinity to the problem
- One idea collector who records the suggested ideas

4.5.4 Create a list of lead questions

During the brainstorm session due to any reason, the creativity may decrease. At this moment, the chairman should stimulate creativity by suggesting a lead question to be answered, such as "can we combine these ideas, or how about a look from another perspective?" It is advised to prepare a list of such lead questions before the session begins.

4.6   Brainstorming warm-ups

Brainstorming warm-ups are useful for getting people into the right frame of mind for a brainstorming session. Word games, a practice run, and a game of opposites are three common brainstorming warm-ups used, which help those involved in the brainstorming process to overcome common brainstorming stumbling blocks and maximize creative results.

4.6.1 Word games

Word games are excellent brainstorming warm-ups. They exercise the mind and help us get into the proper mindset for the actual brainstorming process. It really doesn't matter which specific word games are used as brainstorming warm-ups, as long as they are mentally stimulating and challenging.

4.6.2 A practice run

Brainstorming on a completely unrelated topic is one of the most popular and productive brainstorming warm-ups put into practice. To put this brainstorming warm-up into action, you create an amusing imaginary problem and then brainstorm ways to overcome it. By brainstorming on a completely unrelated topic, you can get a feel of the brainstorming process and begin to warm up and exercise the parts of the brain which will be put to work during the actual brainstorming session.

4.6.3 A game of opposites

It is useful to break down creative barriers prior to the brainstorming session, and use warm-ups such as opposite games to accomplish this task. To perform this brainstorming warm-up, write down a list of ten to twenty common words. Next to each word, write down the first three words that come to mind when you think of what the opposite of that word should be. If this is a group brainstorming session, have one person read each of the words aloud while all members of the brainstorming team write down the first three words that come to mind.

By using these brainstorming warm-ups prior to your brainstorming session, the creative process will flow more freely, the ideas will come more quickly, and the results of your brainstorming efforts can be maximized.

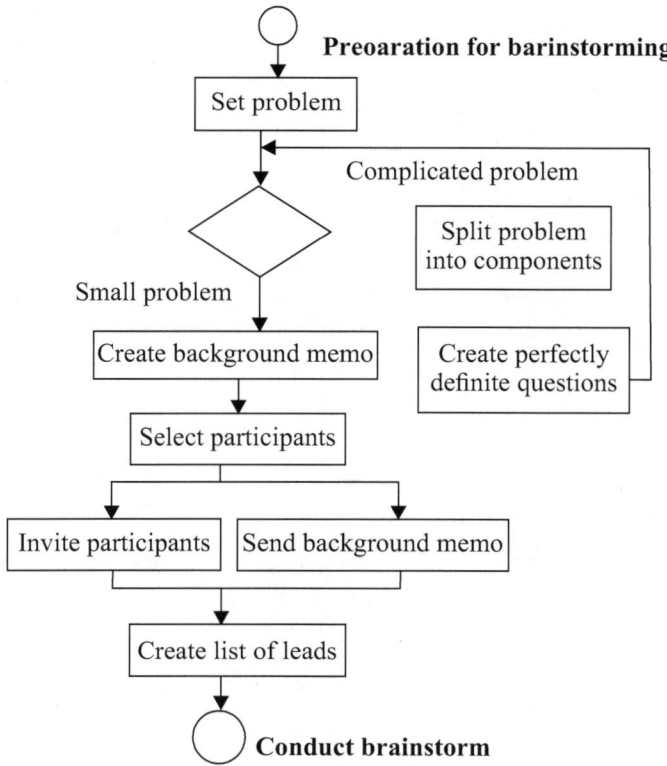

Figure 4.3 Preparation for brainstorming.

4.7 Osborn's checklist for adding new ideas

A basic rule of brainstorming is to build onto ideas already suggested. Alex Osborn, the originator of classical brainstorming, first communicated this. A checklist was formulated as a means of transforming an existing idea into a new one. The checklist is designed to have a flexible trial and error type of approach. As a simple tool to support concept generation, Osborn's checklist is a comprehensive list of questions about ideas and problems which can be used either individually or in groups. This aims to encourage creativity and divergence in concept generation. A derivation of Osborn's checklist is called as SCAMPER.

It is a series of simple questions that can be used either individually or in groups, designed to support creative and divergent thinking when faced by a design problem. The questions need a point of focus which could either be an existing solution or proposed concepts to a design problem. The questions should be taken one at a time, to explore new ways and approaches to the problem.

Other uses?	New ways to use it? Other uses if modified? Put to other uses? As it is? If modified?
Adapt?	What else is like this? What other idea does this suggest? Does past offer parallel? What could I copy? Whom could I emulate?
Modify?	New twist? Change meaning, color, motion, odor, taste, form, shape? Other changes? Give it a new angle?
Magnify?	Can anything be added, time, frequency, height, length, strength? Can it be duplicated, multiplied, or exaggerated? What to add? More time? Greater frequency? Stronger? Higher? Larger? Longer? Thicker? Heavier? Extra value? Plus ingredient? Duplicate? Multiply? Exaggerate?
Minify?	Can anything be taken away? Made smaller? Lowered? Shortened? Lightened? Omitted? Broken up? What to subtract? Condensed? Miniature? Lower? Narrower? Streamline? Understate? Less frequent?
Substitute?	Different ingredients used? Other material? Other processes? Other place? Other approach? Other tone of voice? Someone else? Who else instead? What else instead? Other power? Other place? Other time?
Rearrange?	Swap components? Alter the pattern, sequence, or layout? Change the pace or schedule? Transpose cause and effect? Interchange components? Other pattern? Other layout? Other sequence? Change place? Change schedule? Earlier? Later?
Reverse?	Transpose positive and negative? How about opposites? Turn it backward, upside down, inside out? Reverse roles? Change shoes? Turn tables? Turn other cheek?
Combine?	Reverse combine? Combine units, purposes, appeals, or ideas? A blend, alloy, or an ensemble?

This list is also sometimes referred to as SCAMPER – Substitute, Combine, Adapt, Modify/Magnify/Minify, Put to other uses, Eliminate, Reverse/ Rearrange.

Osborn's checklist involves thinking of ideas to alter the problem statement by:

- adapting it – How can it be used as it is, and what other uses could it be adapted to?

- modifying it – How can we change the meaning, material, shape, color, smell, etc.?

- magnifying it – How can we make it bigger, stronger, or add something new to it?

- minifying it – How can we make it smaller, split it apart, remove something, make it smaller?

- substituting with it – How can we interchange parts, use other ingredients, materials, etc.?

- rearranging it – How can we use other layouts, reverse positives/ negatives, cause/effects, or reverse roles?

- combining it – How can we combine parts, units, or ideas? Can we blend, compromise, or combine different ideas?

4.7.1 Conducting session

The leader of the brainstorming session ensures that the basic rules are followed. The activities of a typical session are as follows:

(1) A warm-up session to expose novice participants to the criticism-free environment. A simple problem is brainstormed; for example, *"what should be the next corporate Diwali/Christmas present?"* or *"what can be improved in the present situation?"*

(2) The leader presents the problem and gives a further explanation, if needed.

(3) The leader asks the brainstorming panel for their ideas.

(4) If ideas are not coming out, the leader suggests a lead to encourage creativity.

(5) Every participant presents his or her idea, and the idea collector records them.

(6) If more than one participant has ideas, the leader lets the most associated idea to be presented first. This selection can be done by looking at the body language of the participants, or just by asking for the most associated idea.

(7) The participants try to elaborate on the idea, to improve the quality.

(8) When the time is up, the leader organizes the ideas based on the topic goal and encourages discussion. Additional ideas may be generated.

(9) Ideas are categorized.

(10) The whole list is reviewed to ensure that everyone understands the ideas. Duplicate ideas and obviously infeasible solutions are removed.

(11) The leader thanks all participants and gives each a token of appreciation.

In a brainstorming session, it can be useful to write each statement on a card and randomly select a card when discussing alternative solutions. Alternatively, paste the questions onto a board and place in the design team's environment.

4.7.2 The process

Participants who have an idea but no possibility to present it are encouraged to write down their ideas and present them later. The idea collector numbers the ideas, so that the leader can use the numbers to encourage quantitative idea generation; for example, *we have 40 ideas now, let's get it to 50!* The idea collector should repeat the idea in the same words he or she has written it, to confirm that it expresses the exact meaning intended by the originator. When more participants are generating ideas, the one with the most associated idea should have the priority. This is to encourage elaboration on previous ideas.

During the brainstorming session, few people do not encourage the attendance of managers and superiors, as it may inhibit and reduce the effect of the four basic rules, especially the generation of unusual ideas.

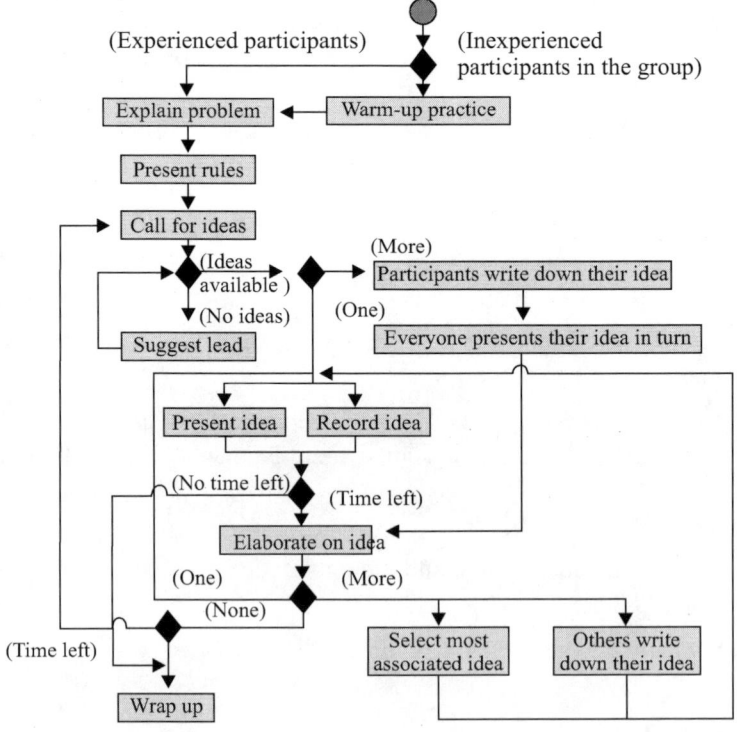

Figure 4.4 Conducting brainstorming.

4.8   Brainstorming tips

Don't reject any ideas, even if they seem too strange. Don't worry about their likelihood at this stage either. The purpose here is to think of ideas, and not to crush them. Keep a positive attitude.

Do you really brainstorm or just "discuss a few ideas"? Real and dynamic brainstorming sessions are easily and unintentionally replaced by "brain drizzling" sessions that leave everyone feeling low. It's just like wet weather; the 10 minute thunderstorm produces buckets of rain, excitement and energy, but 2 hours of drizzle dampens everyone's spirit. Negative thinking and comments will often put a damper on the brainstorming session, resulting in what Folger and LeBlanc termed as "brain drizzling".

By contributing literally the first thought that comes into our heads, no matter how silly or disconnected what we say appears to be, we extend our creative thinking process. The creative thinking becomes a chain reaction; one person's comment triggers an instinctive reaction in another; they call out their train of thoughts, which prompts others to come up with comments along another track, which in turn triggers further reactions, and so on.

Seven brainstorming rules given by Lindsay Swinton are as follows:

Rule 1 – No idea too stupid: There is an ideal solution to a problem, and brainstorming is the key to find it. However, discussing, criticizing, or generally dismissing ideas as they come up reduces your chance of finding the secret treasure and renders your brainstorming session useless.

Rule 2 – Watch the clock: A little time pressure is good for brainstorming, so decide maximum time for brainstorming, say 10–20 min, and stick to it. Start and finish on time, and encourage a brisk pace to maximize the time invested in this activity. Maybe assign a time-keeper for this task.

Rule 3 – Record your progress: All your good ideas are wasted hot air if they are not recorded methodically and more importantly, legibly. Consider using brainstorming software such as mind manager, post-it notes, flip charts, or other such methods for writing down your ideas. Whatever you choose, make sure you bring all the necessary tools and materials with you!

Rule 4 – Quantity not quality: The aim of brainstorming is to churn out as many ideas as you have time for before you do any reality check on their merits. Through quantity you will find quality, even though it might take some time and effort to get there. Ideas breed ideas.

Rule 5 – Use both sides of your brain: Most work activities use your left side of the brain, so make your right side of the brain do some work for a change and get more from brainstorming. Use colored or scented pens, random props, or anything that says "creative and fun" and not "stuffy and staid".

Rule 6 – Encourage the right mindset and have fun: Consider using an ice-breaker or creativity exercise to get group members into the right frame of mind and away from creativity blocking thoughts of unanswered emails, to-do lists, and other priorities. And once brainstorming has started, remember that the performance anxiety will dry up creative juices quicker than a quick thing, so make sure the atmosphere is kept light and soft, and above all, fun.

Rule 7 – Let no good idea go unheard: Not everyone enjoys brainstorming and group problem solving. Shyness, fear of looking stupid or silly may keep people quiet. Brush up on your facilitation skills and avoid the risk of great ideas being un-spoken or unheard.

Group problem solving can be effective, especially if you follow these 7 brainstorming rules and techniques. You can design a horse!

4.9   Variations in brainstorming

There are different types of brain storming. The most popular among all of them are normal group technique, group passing technique, team idea-mapping method, and electronic brainstorming. You can try others also if found convenient to your working culture. Let us have a glance on few of these techniques.

4.9.1 Nominal group technique

Normal group technique is a more controlled variant of brainstorming used in problem-solving sessions to encourage creative thinking, without group interaction at idea-generation stage. Each member of the group writes down his or her ideas that are then discussed and prioritized one by one by the group. This is also called as nominal group process.

The nominal group technique is a type of brainstorming that encourages all participants to have an equal say in the process. It is also used to generate a ranked list of ideas.

Participants are asked to write down their ideas anonymously. Then the moderator collects the ideas, and each is voted on by the group. The vote can be as simple as a show of hands in favor of a given idea. This process is called distillation. The steps involved are as follows:

- The problem, situation, or question is stated clearly and concisely.
- The coordinator asks participants to generate a list of the features or characteristics of the problem or question.
- The coordinator gives the group 5–15 min to work silently.

- Each suggestion is recorded on a chart visible to all members.
- Members clarify the items, but do not yet evaluate them.
- Each person chooses his or her top ranked items.
- The group engages in full discussion about the top rated items.
- A decision is reached.

After distillation, the top ranked ideas may be sent back to the group or to subgroups for further brainstorming. For example, one group may work on the shade or color combination required in a product, whereas another group may work on the basic fabric construction like count, ends and picks, weave, design, draft and peg plan, width, and so forth. Each group will come back to the whole group for ranking the listed ideas. Sometimes ideas that were previously dropped may be brought forward again, once the group has re-evaluated the ideas.

It is important for the moderator to have received training in this process before attempting to take on the moderating task. The group should be primed and encouraged to embrace the process. Like all team efforts, it may take a few practice sessions to train the team in the method before working on the important ideas.

The characteristics of a nominal group are as follows:

- Capitalizes on the finding that people working individually in the presence of others sometimes generate more ideas than those interacting as a group.
- Can enable members to reach a decision on a controversial issue without leaving a residue of bitterness from a win-lose conflict.
- Members work individually in each other's presence by writing their ideas. They record these ideas on a chart, discuss them as a group, and finally evaluate them by a ranking procedure until members reach a decision.
- Stifles effect of dominating members of the group.
- Tendency for lazy members to let others carry the ball is minimized.
- Adds structure to the brainstorming process.

4.9.2 Group passing technique

Each person in a circular group writes down one idea, and then passes the piece of paper to the next person in a clockwise direction, who adds some thoughts. This is repeated until everybody gets their original piece of paper back. By this time, it is likely that the group will have extensively elaborated on each idea.

A popular alternative to this technique is to create an "idea book" and post a distribution list or routing slip to the front of the book. On the inside cover (or first page) will be a description of the problem. The first person to receive the book lists his or her ideas and then routes the book to the next person on the distribution list. The second person can log new ideas or add to the ideas of the previous person. This continues until the distribution list is exhausted. Follow-up "read out" meeting is then held to discuss the ideas logged in the book. This technique takes longer, but allows individual thought whenever the person has time to think deeply about the problem.

4.9.3 Team idea-mapping method

This method of brainstorming works by the method of association. It may improve collaboration and increase the quantity of ideas, and is designed so that all attendees participate and no ideas are rejected. The process begins with a well-defined topic. Each participant creates an individual brainstorm around the topic. All the ideas are then merged into one large idea map. During this consolidation phase, the participants may discover a common understanding of the issues as they share the meanings behind their ideas. As the sharing takes place, new ideas may arise by the association. Those ideas are added to the map as well. Then ideas are generated on both individual and group levels. Once all the ideas are captured, the group can prioritize and/or take action.

4.9.4 Electronic brainstorming

Electronic brainstorming can be done via email. The leader or facilitator sends the question out to group members, and they contribute independently by sending their ideas directly back to the facilitator. The facilitator then compiles a list of ideas and sends it back to the group for further feedback. Electronic brainstorming eliminates many of the problems of standard brainstorming, such as production blocking and evaluation apprehension. An additional advantage of this method is that all ideas can be archived electronically in their original form, and then retrieved later for further thought and discussion. Electronic brainstorming also enables much larger groups to brainstorm on a topic that would normally be productive in a traditional brainstorming session.

Other forms of electronic brainstorming could be browser based, client/server, or peer to peer related software.

The shortcoming of electronic brainstorming is that the ideas do not click in the minds of other people after seeing the idea given by someone else, which is normally found when people physically assemble and give ideas. Further this technique does not help in team building, as the members are not meeting each other personally.

4.9.5 Random stimulation

This involves picking a completely random word or idea and using it to generate a better and more applicable idea. This word could be pulled out of a hat, a prepared list, or even opening up a dictionary to a random page and picking the first word from there. This technique helps in naming a new brand, devising sales promotion plans, arranging for visual merchandising of fabrics and garments, designing a new style, and so on. This technique can also be used for conducting team-building exercises among staff and workers in a mill or garment factory, and involving people in company's activities.

4.9.6 Free writing

Free writing lets one's thoughts flow as they will, putting pen to paper and writing down whatever comes into mind. One need not judge the quality of what is written and about style or any surface-level issues like spelling, grammar, or punctuation. If one can't think of what to say, he can write that down really. The advantage of this technique is that you free your internal critic and allow yourself to write things you might not write if you were being too self-conscious.

While free writing, one can set a time limit ("I'll write for 15 minutes!") and even use a kitchen timer or alarm clock or can set a space limit ("I'll write until I fill four full notebook pages, no matter what interrupts me!") and just write until he reaches that goal. One might do this on the computer or on paper, and can even try it with eyes shut or the monitor off, which encourages speed and freedom of thought. One can get few wonderful ideas, out of which few shall be real gems. Extraordinarily good literature takes birth like this, and can become an inspiration for others to do the work, including building a nation or fighting a war.

4.9.7 Other people's views

This is where one tries to put himself in others' shoes and analyze the problem differently. This is very essential when we are trying to understand the complaints given by customers. In a number of cases it is seen that the quality assurance people turn down a customer complaint by saying that the product was tested and they also conformed to all specified requirements. We need to understand the way in which a customer looks at the problem. He may be looking at it in a different way, and his requirement may be different.

Let us take an example:

An overseas customer asked for Ne 20 KHC. The mill had Ne 20s combed yarn in stock, which was rejected by the quality assurance because of higher neps and imperfections. The yarn was unfit to be sold as a combed yarn. The management decided to divert that lot by putting the labels of carded yarn, and felt that customer shall be very happy as he got much superior yarn than his expectations. The first feedback on the yarn was that the working of knitting machine was extremely good, as compared to the normal carded yarn the customer was using. However, after the material was converted into T shirts and reached the show rooms, there was no sale for this as the ultimate customers rejected it by saying that it has "feminine feel" and they wanted "masculine feel on T shirts".

4.9.8 Futuring

Futuring is where solutions that are not do-able now, but someday in the near or far future might get used are considered. This is more useful in fashion designing for the next season. This also considers the technical developments taken place and the likely problems are imagined with new developments and a preventive solution is provided in advance.

4.9.9 Breaking and bulleting

(1) *Break down the topic into levels*: Once you have a course assignment in front of you, you might brainstorm:

- the general topic

- a specific subtopic or required question

- a single term or phrase that you sense you're overusing in the paper

(2) *Listing/Bulleting*: Jot down lists of words or phrases under a particular topic. You can try this one by basing your list:

- on the general topic

- on one or more words from your particular thesis claim

- on a word or idea that is the complete opposite of your original word or idea.

The first list might be based on a general topic, and the second on your thesis; the next list might be based on the opposite claim. You could do number of lists and then compare the evidence from all.

Using multiple lists will help you to gather more perspective on the topic and ensure that, sure enough, your thesis is solid as a rock, or can explain that your thesis is full of holes and you'd better alter your claim to one you can prove.

4.9.10 Cubing brainstorming

Cubing enables one to consider the topic from six different directions; just as a cube is six-sided, similarly cubing brainstorming will result in six "sides" or approaches to the topic. Take a sheet of paper, consider your topic, and respond to these six commands.

1. Describe it.

2. Compare it.

3. Associate it.

4. Analyze it.

5. Apply it.

6. Argue for and against it.

Look over what you've written.

- Do any of the responses suggest anything new about your topic?

- What interactions do you notice among the "sides"?

- That is, do you see patterns repeating or a theme emerging that you could use to approach the topic or draft a thesis?

- Does one side seem particularly fruitful in getting your brain moving?

- Could that one side help you draft your thesis statement?

Use this technique in a way that serves your topic. It should, at least, give you a broader awareness of the topic's complexities; if not, a sharper focus on what you will do with it.

4.9.11 Similes

In this technique, comparisons are made between two or more statements or substances. The technique involves writing sentences. For example, complete the following sentence:

_____ is/was/are/were like _____.

In the first blank, put one of the terms or concepts your paper centers on. Then try to brainstorm as many answers as possible for the second blank by writing them down as you come up with them. After you have produced a list of options, look over your ideas. What kinds of ideas come forward? What patterns or associations do you find?

4.9.12 Clustering/ Mapping/ Webbing

This technique has three (or more) different names, according to the activity or what the end product looks like. One will write a lot of different terms and phrases onto a sheet of paper or on a black board in a random fashion and later go back to link the words together into a sort of "map" or "web" that forms groups from the separate parts. Allow yourself to start with chaos. After the chaos subsides, you will be able to create some order out of it. The steps followed are as follows:

1. Take a sheet(s) of paper and write your main topic in the center, using two or three words.

2. Move out from the center and fill in the open space any way you are driven to fill it, start to write down, fast, as many related concepts or terms as you can associate with the central topic. Jot them quickly, move into another space, jot few more down, move to another blank, and just keep moving around and jotting. If you run out of similar concepts, jot down opposites, jot down things that are only slightly related, or jot down anything say: your grandpa's name, but try to keep moving and associating. Very important is not to worry about the (lack of) sense of what you write, for you can choose to keep or toss out these ideas when the activity is over.

3. Once the storm has subsided and you are faced with a hail of terms and phrases, you can start to cluster. Circle terms that seem related, and then draw a line connecting the circles. Find few more and circle them and draw more lines to connect them with what you think is closely related. When you run out of terms that associate, start with another term. Look for concepts and terms that might relate to that term. Circle them and then link them with a connecting line. Continue this process until you have found all the associated terms. Few of the terms might end up uncircled, but these "loners" can also be useful to you. (Note: One can use different colored pens/pencils/chalk for this part for easy identification. If that's not possible, one can vary the kind of line used to encircle the topics; like a wavy line, a straight line, a dashed line, a dotted line, a zigzag line, etc., in order to see what goes with what.)

4. When one stands back and surveys the work, sees a set of clusters, or a big web, or a sort of map: hence the names for this activity. At this point, you can start to form conclusions about how to approach your topic. Start with an example or two in order to illustrate how you might form some logical relationships between the clusters and loners you've decided to keep. At the end of the day, what you do with the particular "map", or "cluster set", or "web" that you produce depends

on what you need. What does this map or web tell you to do? Explore an option or two and get your draft going!

4.9.13 Relationship between the parts

In this technique, attempt is made to understand the relation between different parts of the concepts or the sub parts. On a paper on opposite margins, the parts and their relating parts are written as shown below:

Whole	Parts
Part	Parts of parts
Part	Parts of parts
Part	Parts of parts

Looking over these four groups of pairs, one can start to fill ideas below each heading and keep going downwards through as many levels as one can. Now, one can look at the various parts that comprise the parts of whole concept and decide what sorts of conclusions can be drawn according to the patterns, or identify lack of patterns if any.

4.9.14 Journalistic questions

In this technique, the "big six" questions (who, what, when, where, why, and how) that journalists rely on to thoroughly research a story are used. Each question word is written on a sheet of paper, leaving space between them. Then, few sentences or phrases in answer are written as they fit the particular topic. One might also answer into a tape recorder.

Then look over the batch of responses and find:

- whether you have more to say about one or two of the questions
- whether the answers for each question are pretty well-balanced in depth and content
- was there any question that you had absolutely no answer for?
- how might this awareness help you to decide how to frame your thesis claim or to organize your paper?
- how might it reveal what you must work on further, doing library research or interviews, or further note-taking?

For example, if your answers reveal that you know a lot more about "where" and "why" something happened, than you know about "what" and "when" could you use this lack of balance to direct your research or to shape your paper? How might you organize your paper so that it emphasizes the known versus the unknown aspects of evidence in the field of study? What else might you do with your results?

4.9.15 Think outside the box

Even when one is working within a particular discipline or area, he can take advantage of experience in other departments or areas. One can use charts, figures, or working models to explain the ideas. Make use of encyclopedias or dictionaries to search for equal words. Refer books from library to understand the solution for a similar problem.

4.9.16 Advantages, limitations, and unique qualities

This is a relatively straightforward idea-evaluation technique; although it can be used in idea generation.

1. Select one of the ideas/possible solutions.

2. Brainstorm as many advantages of this as you can.

3. Once you have got to a limit on advantages, try and brainstorm all the disadvantages.

4. Swap mindset again to try and find all the unique, new, or unusual qualities about this idea/solution.

4.9.17 Brain sketching

This technique (VanGundy, Techniques of Structured Problem Solving, 1988) is a brain-writing technique that passes evolving sketches, rather than growing written lists of ideas around the group. As usual with most brain-writing techniques, only limited facilitation skill is needed.

(1) A group of 4–8 people sit around a table, or in a circle of chairs. They need to be seated far enough from each other to have some privacy. The problem statement is agreed, and discussed until understood.

(2) Each participant privately draws one or more sketches (each on separate sheets of paper) of how it might be solved, passing each sketch on to the person on their right when it is finished. The facilitator suggests that the sketches should not take more than 5 min or so to draw.

(3) Participants take the sketches passed on to them, and either develop or annotate them, or use them to stimulate new sketches of their own, passing the amended original and/or any new sketches on to their neighbor when ready.

(4) After the process has been running for a suitable period and/or energy is running lower, the sketches are collected.

(5) It will probably help to display all the sketches and to discuss them in
turn for clarification and comment.

Then move on to any appropriate categorization, evaluation, and
selection process.

4.9.18 Brainwriting

Brainwriting is a technique similar to brainstorming and trigger sessions. There
are many varieties, but the general process is that all ideas are recorded by the
individual who thought of them. They are then passed on to the next person
who uses them as a trigger for their own ideas. Examples of this include:

Brainwriting pool: Each person, using post-it notes or small cards, writes
down his ideas and places them in the centre of the table. Everyone is free to
pull out one or more of these ideas for inspiration. Team members can create
new ideas, variations, or piggyback on existing ideas.

Brainwriting 6-3-5: The name brainwriting 6-3-5 comes from the process
of having 6 people write 3 ideas in 5 minutes. Each person has a blank 6-3-5
worksheet as shown below:

Problem Statement: How to...		
	Idea 1	Idea 2	Idea 3
1			
2			
3			
4			
5			
6			

Everyone writes the problem statement at the top of their worksheet (word
for word from an agreed problem definition). They then write 3 ideas on the
top row of the worksheet in 5 minutes in a complete and concise sentence (6–
10 words). At the end of 5 minutes (or when everyone has finished writing),
pass the worksheet to the person on your right. You then add three more ideas.
The process continues until the worksheet is completed.

There will now be a total of 108 ideas on the six worksheets. These can
now be assessed.

Idea card method: Each person, using post-it notes or small cards, writes
down ideas, and places them next to the person on his or her right. Each
person draws a card from there neighbors pile as needed for inspiration. Once
the idea has been used, it is passed on to the person on the right along with any
new, variations, or piggybacked ideas.

Brainwriting game: This method is set in the form of a light-hearted competitive game. Creativity methods normally avoid competition because it tends to be divisive. However, as long as the gaming atmosphere is fun, rather than overly competitive, and the facilitator ensures that there are no significant losers, the game format might be useful, particularly in training contexts where winning and losing are likely to be less of an issue and both can be used to provide teaching material.

The game will take a little longer than few other brainwriting techniques. Very little facilitation skill is needed. The structure is as follows:

1.   Display the problem statement, and explain that the winner of the game is the one who devises the most unlikely solution.

2.   The facilitator sells each group member an agreed number (say 10) of blank, serially numbered cards, at say 10p each, pooling the money to form the prize. Each group member signs a receipt that records the serial numbers of their set of cards.

3.   Members try to think of utterly impossible solutions, writing one per card. The cards are then put up on a display board.

4.   Members now have (say) 15 min to silently read all the solutions, and to append to them (on further un-numbered cards or post-its) the ways in which they could be converted into a more practical way of solving the problem (so reducing that idea's chances of winning).

5.   Each member then has two votes (e.g., two sticky stars) to vote for what she/he now considers to be the most improbable idea on the numbered cards. The idea that attracts most votes wins the pooled money.

6.   Form two sub-groups, give half the cards to each and give each group (say) 15 min to develop six viable solutions from their cards.

7.   Each sub-group tries to 'sell' their ideas to the other sub-group.

8.   Everyone comes together and agrees on the best ideas overall.

Constrained brain writing: On a number of occasions one may want constrained ideas around pre-determined focus, rather than ranging freely. The versions described here use the standard brainwriting-pool technique, but bias the idea generation by using brain-writing sheets prepared in advance.

(1)   Present starter ideas: The leader initiates the process by placing several prepared sheets of paper in the pool in the centre of the table.

(2)   Private brainwriting: Each group member takes a sheet, reads it, and silently adds his or her ideas.

(3)   Change sheet: When a member runs out of ideas or wants to have the stimulation of another's ideas, she/he puts one list back in the centre of

the table and takes one returned by another member. After reviewing this new list, she/he adds more ideas.

(4) Repeat until ideas are exhausted: No discussion at any stage.

Varying the level of constraint:

Cued brainwriting: For mild constraint, the sheets are simply primed with one or more starting ideas (e.g., SWOT's, issues) in the required area.

Structured brainwriting: Structured brainwriting can be used for a stronger constraint. The sheets can be formally headed, each sheet relating to a particular issue or theme. The participants are asked to keep the ideas they contribute on each sheet relevant to the issue in the heading on that sheet.

Card story boards are another way of directing idea generation.

4.9.19 Bullet proofing

The bullet proofing technique aims to identify the areas in which your plan might be especially vulnerable:

- What may possibly go wrong?
- What are few of the difficulties that could occur?
- What's the worst imaginable thing that could occur?

There are few similarities between Potential Problem Analysis (PPA) (Kepner and Tregoe) and negative brainstorming (Isaksen and Treffinger, 1985), who suggest that "what might happen if…?" is a useful question to be used for looking at potential challenges.

- Brainstorm around enquiries such as: "What might happen if…?" to identify the areas in your plan of action that could potentially cause problems and which have not yet been identified.

- All the areas identified should be placed on a table such as the one below, showing how likely the event is to occur; and if it did occur, how important it would be for your plan.

- Major problems are very likely to happen. If there are significant numbers, you may first need to prioritize them so that you can focus your efforts on the most important ones.

- Use any suitable problem-solving method to work out the ways to deal with them.

	How likely is it to occur	
	Unlikely	Very likely

If it did occur, it would be:	Major problem		Most serious
	Minor problem	Least serious	

Although this type of exercise is necessary, it can have the effect of lowering your spirits by looking at the "black side". Should this be the case and you feel the need for some cheering up, try using the same technique in reverse:

- What could go well?

- What pleasant surprises might it deliver?

- What is the best thing that could happen?

Obviously, these uplifting enquiries should be reasonably plausible – a collection of good things that really might happen!

4.9.20 Double Entry A-ha! Method (DEAM)

This system developed by Dr. Win Wenger in October 2005 is a simple tool for generating ideas. DEAM is a part of the 'Evoked Sidebands' method for problem-solving and for building understanding. Evoked Sidebands was created in March 2005. The "Double-Entry" part of the title, the "DE" in DEAM, refers to the method's procedure of writing on two pieces of paper at the same time. On one sheet the topic-to-be-understood is written in brief i.e., question or problem, where whatever thoughts and perceptions noticed during the process of problem-writing are recorded on the second sheet of paper. Often the second sheet shall be filled with comments and observations before completely writing a single sentence of problem-statement on the first sheet. The thoughts, perceptions, and observations had been there all along; they just hadn't been noticed before. Deciding to notice them, and giving priority to them in writing, and the act of writing them onto second sheet — bring them into view. One shall be amazed at the number of ideas he had, and how many of those ideas are relevant and even good, once one proceeds to write down everything that comes in his mind. DEAM is a powerful tool for understanding almost any situation quickly, easily, and in meaningful, useful depth, and as a powerful learning tool, for any topic or subject where understanding is needed. The logic behind this tool is that you have a lot more going on in your mind than in your conscious thoughts — a lot more even in your immediate physical sight than you consciously notice. You have literally hundreds of lines of thought, and streams of perception going on in your mind at the same time. Ordinarily, though, you never get around to noticing them, however important few of them may be to you.

4.9.21 Buzz groups

The characteristics of a buzz group are that a large group is subdivided into smaller groups which discuss an assigned target question, then report their questions back to the main group, which encourages participation and their involvement that is not feasible in large groups. Technique can be used to identify problems or issues, generate questions to study, compile a list of ideas or solutions, or stimulate personal involvement. The steps involved in buzz group are as follows:

- The facilitator presents a target question to the group.

- If the group is large, divide into smaller groups (approximately six people).

- Each group is given a copy of the target questions on an index card and a recorder/spokesperson is selected by seating. The individual then writes all ideas on index cards.

- The group spends a few minutes thinking of and evaluating the ideas.

- The group reports its list to the entire assembly.

4.9.22 Delphi methods

This method is good to be utilized when time and distance constraints make it difficult for group members to meet. The steps are as follows:

- A Delphi panel is selected by the facilitator.

- The problem or issue is stated concisely in writing and sent to each of the Delphi panel for individual work.

- The facilitator compiles another document that details all the individual positions taken by the panel and distributes a copy to each member.

- This procedure, with a facilitator compiling the individual comments into a single document and distributing it to the group, continues until a consensus is reached.

This is not a group decision technique. It involves the presentation of a problem or an issue to the appropriate individuals, asking them to list their solutions, compiling a master list, circulating this master list to all participants, and asks them to comment in writing on each item on the list. The list with comments is then circulated to the participants. The procedure is continued until a decision is reached.

4.9.23 Fantasy chaining

Whenever the group is not talking about the here-and-now of the problem, it is engaged in fantasy. Fantasy chaining is a group story-telling method wherein everyone in the group adds something to the topic at hand, which may not necessarily be the primary focus.

- Manifest theme is what the fantasy chain is about at the surface level.

- Latent theme is the underlying theme (what the group members are really thinking about).

- Helps the group define itself by creating symbols that are meaningful and that help determine its values.

- Enables a group to indirectly discuss the matters which might be too painful or difficult to bring out into the open.

- Helps a group deal with emotionally "heavy" information.

- Effective way in which groups create their shared images of the world, each other, and what they are as a group.

- A group's identity converges through these shared fantasies.

4.9.24 Focus groups

This is a group focusing on a specific subject. It is making it work for you behind a one-way mirror. Instructor introduces a topic that is to be discussed by the group in any way they choose. Its characteristics are as follows:

- Encourages unstructured thoughts about a given topic.

- Often used to analyze people's interests and values.

- Universities, large corporations, and political candidates use focus groups to understand how others perceive their strengths and weaknesses.

4.9.25 Metaphorical thinking

A metaphor is a thinking technique connecting two different universes of meaning. The key to metaphorical thinking is similarity. Excessive logical thinking can stifle the creative process, so use metaphors as a way of thinking differently about something. Make and look at metaphors in your thinking, and be aware of the metaphors you use. Metaphors are wonderful, so long as we remember that they don't constitute a means of proof. As by definition, a metaphor must break down at some point. The steps involved are as follows:

- State the objectives of thinking in metaphors: to see comparisons between two ideas, and to gain new insights from comparisons.

- Brainstorm possible metaphors for few aspects of the problem.

- "Piggyback" on metaphors; build on them.

- Choose the best metaphors to carry further.

- Examine all imaginable areas of comparison in the metaphor.

- Ask questions which the metaphor might answer.

- Look for insights into causes, effects, and solutions for your problem.

- When identifying possible solutions, it is essential to:

- Hold back from evaluating proposed solutions.

- Make a point of "thinking outside" of your own experience and expertise.

- Involve everyone in the process.

- Go for quantity – at least 20 or so possible solutions before narrowing the list to between four and six of the best suggestions.

4.9.26 Evoked sidebands

Evoked sidebands is a technique developed by Dr. Win Wenger of Project Renaissance, Gaithersburg in 2005, for understanding the problem. It is a very simple method and can be practiced by anyone. Focus of awareness is the basic point on which the normal thoughts flow. For example, hold your index finger up at arm's length distance and stare at it without moving your eyes from it for a minute or so. Notice how much of the rest of the space you can see around and understand without moving your eyes from your finger. Your focus of awareness is about as wide, metaphorically speaking, as your visual focus, which in fact is about as wide as your index finger held up at arm's length distance. Over a lifetime, you've accumulated a habitual strategy for what you include in that focus and what you exclude from it.

Most of the time that strategy serves you well enough, but rest of the time what you've excluded from the focus of your attention is much more valuable to you than what you've included. Each time you are stuck on a question or problem, you can pretty well bet that is the case; likewise for whenever you've been unable to fully grasp something that you've been trying to understand. Evoked sidebands is an easy and direct way to bring into focus some of your best thinking and perceiving.

The method is simple. We need at least two sheets of paper to write on, or one sheet and an audio recorder, or one sheet and a live listener. With the intention and expectation of noticing thoughts and perceptions as they occur just beyond the focus of attention, we need to move as follows:

(1) Decide upon and write out in one, two, or three short sentences the matter you wish to understand better or the problem you'd very much like to get good answer or solution to.

  • While you are writing the matter, question, or problem-statement out in that short sentence or so, be alert to other things occurring to you besides what you are writing. As they occur, either write them on the second sheet of paper or report them into your audio recorder, or report them to your listener.

(2) Write out the same sentences again, even the same words, and this time not in your handwriting but rather in block print.

  • While you are writing that matter in block print, notice new things occurring to you besides what you are writing and besides what occurred to you before. As they occur, write them on the second sheet of paper, or report them to recorder or listener.

(3) Write out the same sentences again in the same words, but this time in an exotic handwriting — if you sloped forward before, slope backward; if sparse before, flowery now — whatever will contrast greatly with your usual handwriting.

  • While you are writing that matter in exotic handwriting, notice new things occurring to you besides what you are writing, and besides what occurred to you before. As they occur, write them on the second sheet of paper, or report them to the recorder or listener.

(4) Stare for 3–4 min at the one handwritten version; notice and report or record still more things occurring to you, still more secondary awareness and associations, which may or may not have much to do with the topic, but you are reporting them anyway.

Stare for 3–4 min at the second handwritten version, and report or record still more things occurring to you; perhaps these secondary awareness and associations evoked for you will have a difference in style from those evoked with the first version.

Then stare for 3–4 min at the third version, report or record still more things occurring to you. How would you compare the style of the things occurring to you with this version, with the style of those occurring to you with the second version and with the first version? You get different wonderful ideas. This is a powerful tool for developing understanding. It is also a power tool, whose practice makes you more and more aware of your own higher intellectual processes and perceptions.

4.9.27 Rawlinson brainstorming

Rawlinson brainstorming is useful variant of brainstorming for untrained groups where there is no interaction between members of the group; all ideas are directed towards the facilitator/scribe. The problem owner simply describes in a headline the problem; he then gives simple background on routes he has tried and has failed, and what would represent an ideal solution. The participants are invited to have a creative warm-up session and then offer solutions to the problem as two word descriptors. The problem owner focuses on those ideas which give him new viewpoints.

Try to involve all the people who have a vested interest in solving the problem, and those who have specialized knowledge and are willing to participate. As the leader of the brainstorming session, you will take the group through four distinct stages. These stages are as follows:

1. Stating the problem

2. Restating the problem

3. Brainstorming on one or more of the restatements

4. Evaluating the ideas produced

    •   If the group members have not used the technique before, then you will need to take them through a preliminary stage to explain the technique and set the scene.

    •   State the problem to the group (without writing it on a flip-chart or whiteboard).

    •   Ask for suggestions as to how the problem may be restated and write these up in front of the group (each restatement should be prefaced by the words "How to..."). By participating in restating the problem in different words a number of times, the group will begin to see different perspectives on the problem. Without this step, it is likely that these perspectives would be overlooked.

    •   Choose one or more of the restatements to brainstorm on. You may want to use a tool such as multi-voting for this.

    •   Begin the main part of the exercise – the brainstorming itself. This calls for a free flow of ideas, aimed at producing as many as possible. Wild ideas are encouraged and the atmosphere should be light-hearted and enjoyable.

- To encourage this, there are four rules to be enforced (explain these beforehand to groups new to the technique, and reiterate them briefly to experienced groups).

  - Suspend judgement: There should be no criticism of other people's ideas; the key is to laugh *with* and not *at* the ideas of other people.

  - Freewheel: Encourage participants to dream or to drift, and to be prepared to produce wild or silly ideas. No idea should be discouraged however off-beat.

  - Quantity: Look for a large volume of ideas –100 ideas in 20 min is not uncommon.

  - Cross-fertilise: Encourage an idea from one member of the group to be developed by other group members.

- Start the ideas flowing by writing up a couple yourselves, if necessary.

- Write everyone's ideas up in front of the group, so that they can be seen and numbered.

- Don't be discouraged if the ideas dry up and group members become frustrated; this usually presages a renewed bout of ideas and only requires a pause for everyone to reflect before starting again.

- When sufficient ideas have been generated, stop and take a breather before going on to the evaluation stage.

- The evaluation stage brings in critical judgement as opposed to the freewheeling atmosphere of the "ideas generation" stage.

- Develop the final list by a process of elimination, that is, by weeding out the least promising ideas progressively. Again, multi-voting may be useful here.

- Use the whiteboard to collect, write up, and refine a final list of ideas which the group sees as most likely to solve the problem.

4.9.28 Role storming

Role storming is an evolution of brainstorming, whereby you take on another identity. Viewing problems and solutions from a different standpoint; what would you do if you were in someone else's shoes? You can also try to imagine what your worst competitor might be doing or thinking about your business. Unusual "off the wall" ideas may seem radical/silly if "you" present them; however, generated by a nameless person removes any embarrassment.

- Use traditional brainstorming or other idea generating technique as a start point.

- Invent an identity or use that of someone else you know.

- Assume that identity or refer to the fictitious person as "this person would suggest..."

- Brainstorm (or use other idea-generating techniques) in separate identity.

Change roles. Now try another identity; obviously this can be done several times for many different characters.

4.9.29 Stimulus analysis

Stimulus analysis is a method whereby digression from the original problem to stimulate alternative ideas may generate an accidental solution. The steps followed are as follows:

- Identify the problem and enter into a discussion.

- Produce a list (10 or more) of arbitrary ideas totally unrelated to the problem.

- Select one of the ideas and discuss in detail all its characteristics.

- Look at each one of these characteristics, and go into finer detail by trying to generate yet more ideas.

- Continue through all 10 original ideas till you have exhausted all further ideas.

- Finally analyze the final (lengthy) list of ideas in any applicable way.

Ideally, all the participants in your brainstorming sessions should know about the creative thinking techniques available to them. After you have introduced everyone to the brainstorming rules, you should tell them that you will be introducing some fresh stimuli into the session by using the advanced brainstorming techniques. You will be asking everyone to brainstorm ideas by using the stimulus as a starting point for ideas. Instead of only using other people's ideas to stimulate them, you will be throwing in original stimuli to spark off new ideas. The stimuli will make it a lot easier to see solutions from a different angle because the starting point will be more original.

You will need a stimulus at the start of the process, and you will need more to restart the process when it dries up. Be careful to make sure people use the actual stimulus whatever it is and you should not give out another stimulus too quickly. The harder it is to use a particular stimulus, the more radical and interesting the solutions could be.

In a warm up session or for a change from the normal situation, you should produce the stimuli quickly one after the other to encourage faster thinking and less analysis, but generally you should let the process finish naturally before giving a new stimulus.

Encourage people to discuss the ideas surrounding the stimulus, but try to stop the discussions that are leading away from the targeted topic. Try out various techniques with your groups. Few people prefer certain techniques, and find they work well for them. Few people are stimulated by words, other by pictures. Experiment with each technique and use those which work well for you. Remember to use the other techniques occasionally, because the change will be good for your thinking.

4.9.30 Successive element integration

Successive element integration generates solutions by gradually developing all ideas into lists of ideas – a form of constructive evaluation, allowing every idea a value (see also the point, Receptivity to Ideas).

- A group of approximately six people individually jot down their own list of ideas for solving a specific problem.

- Two members of each group read out one of their ideas, while the remaining members try to integrate the two offered ideas into a third idea (this is added to the overall list).

- A third member of the group offers an idea which is integrated by the other members of the group with the previous ideas to create a fourth idea. This stage is repeated until all the ideas are exhausted and detailed on the overall list.

- In the latter stages of idea generation, the "best of ideas" can be integrated with each other to create a list of exceptional ideas.

The advantages of this method are as follows:

- The skill of building upon other people's ideas
- Encourages constructive convergence
- Ensures that all ideas are carefully considered

4.9.31 Nine variations of brainstorming

Nine different formats for brainstorming, each one with its own unique characteristics are given by Hallhouston in the article A – Teacher in Taoyuan in My coated.

Brainwriting – Participants write ideas on paper slips and then pass the slips to others who can add comments. This is ideal for classes that prefer to discuss through writing.

Brain walking – It is similar to brainwriting. But in this case, students write on large sheets of paper covering the walls. Each sheet of paper has a topic related to the problem, and students can walk around and add comments. This one is highly suitable for students who don't want to spend the whole class sitting down.

Imaginary brainstorming – Here, a problem statement is created and a traditional brainstorming session is given. Everyone suggests changes to few of the words to create a new problem statement; ideally one that is off-the-wall and bizarre. Brainstorming is done again and a list of solutions is made and applied to original problem.

Rawlinson brainstorming – Unlike most ideas here, this one does not emphasize group interaction. One person presents his or her problem and the ideal situation he or she is looking for. Other group members present their solutions directly to the presenter in two-word phrases. The presenter focuses on the ideas that he or she finds most helpful.

Visual brainstorming – Group members sketch solutions to a problem. The sketches are used as a springboard for more solutions. This variation will appeal to visual learners, as well as learners with artistic inclinations.

Negative brainstorming – Participants begin with a problem statement that is the opposite of their goal (e.g., how can we go out of business, or how can we make our workplace more depressing?). They brainstorm a number of ideas. Ultimately, they use their ideas as a springboard for more realistic and useful solutions to their actual problem statement.

Didactic brainstorming – Begin with a question that is an abstract version of your problem statement (e.g., what is quality from the point of view of a girl purchasing dress material?). Get participants to discuss for a few minutes, and come up with a variety of answers. Then reveal your true problem statement (e.g., how can we make customers to buy what we can produce?). This approach might appeal to more philosophical learners.

Role storming – Create a set of roles for a role play that represents a problem statement. Ask students to perform their role plays in groups. Next, students will write down all the solutions that came to their minds as they watched and performed in the role plays.

Value brainstorming – Ask the group to make a list of primary concerns regarding their problem statement. Then ask them to make a list of some of the hidden values behind these concerns. Participants should rank these values and clarify their meaning. Finally, the group should suggest solutions based on these values.

4.9.32 Receptivity to ideas

This technique suggests that you should turn around your traditional way of approaching ideas offered by other people which may initially seem "half baked", "off the wall", or naive. This method recommends that you should be more receptive to such ideas, as they could contain the seed of a "prize" idea.

This thought process is particularly relevant when responding to non-experts, whilst it is accepted that they do not understand the area they are talking about; similarly, they are not indoctrinated by conventional wisdom about "what can't be done". Harriman (1988) describes two synectics techniques to improve receptivity. They are as follows:

Paraphrasing: Once the speaker has offered his thoughts, repeat them back to him by using your own words. But keep them as close as possible to the essence of their idea; for instance you could say, "if I understand this you are suggesting that..." Do not evaluate or give an opinion on his thoughts; you are trying to establish a mutual starting point and understanding, evaluation comes at a later stage.

If the speaker agrees on what you have repeated, then you can move swiftly on to the next stage. However if this is not the case, get the speaker to explain further and try again saying something like "ok, let me try again, am I right in saying that the core of your idea is that..." Continue with this paraphrasing until the speaker confirms your understanding. This stage is essential because it double checks your understanding of what is being suggested, but more subtly shows that you are interested in what the speaker is saying.

Developmental response: After the paraphrasing, you need to work towards transforming the idea into a workable solution. Divide your response into positive elements (pros), and negative elements (cons):

- *Pros* should be precise and genuine; listing at least one more pro than the one coming easily. Often a valuable avenue of thought is opened by that last, hard-to-give pro. This acknowledges the contribution of the speaker and creates better understanding of the problem's components.

- *Cons* should be looked one at a time, phrasing each one so that it encourages solutions; start with "how to", while redirecting discussion toward solving the problem. For example if the con is "its expensive", try saying "how can we make it less expensive?" As you consider each con in turn, correcting it will transform the original idea. The final solution may barely resemble the original thought.

A developmental response centers the attention on the parts of the idea to be preserved, those ideas often overlooked in the initial rush to identify imperfections. It is a process of transformation, going from constructing fresh ideas into ultimate concepts and motivating participants along the way. It expresses a manager's intention to resolve the problem and aims discussion to what needs to be accomplished, dismissing nobody in the process.

4.9.33 Synectics

Synectics is based on a simple concept for problem solving and creative thinking – you need to generate and evaluate ideas. Whilst this may be stating the obvious, the methods used to perform these two tasks are extremely powerful.

Preliminary planning: In advance, hold a preliminary planning meeting with the problem owner(s). This checks that there are genuine problem owners who want new options which they themselves can implement, within their authority; helps you to understand the problem-owner's perceptions of the problem area; gives a feel for the number and quality of solutions needed; helps to ensure realistic expectations about results, and allows you to agree for team membership.

Procedure during the session:

(1) Problem owner provides headline and wish: They describe the issue, how it is experienced, the background, what has been tried, and the possible scope of action. It is then expressed in one or more 'big wish' statements of the form: 'I wish (IW)...' or 'How to (H2) ...' NB that this is not a 'problem definition', but a wish reflecting the way the issue is experienced. The group listens imaginatively, rather than analytically.

(2) Group generates large numbers of "springboards": The mood here needs be expansive and unconstrained. The springboards use the same formats as the 'big wish' (IW, H2, etc.). They are not ideas for solutions but articulate further wishes to open up space for invention (It would be nice if we could do X but we don't yet know how to...) A wide range of springboard triggering techniques have been developed; e.g., various uses of analogies, various types of excursions, the essential paradox/book title technique, and others (free association, random stimuli, drawing techniques, etc.)

(3) Select an interesting springboard: The mood now switches to a more focused approach than in steps 1–2. Problem owner and group members choose their favorite springboards (more on the basis of interest or appeal than on the basis of logical relevance). They share their choices, but final choice rests with the problem-owner. However the process can always be repeated, so the choice is not critical. The assumption is that within any springboard there will be creative possibilities that can usefully be explored.

(4) Ideas to help achieve the selected springboard are generated using the trigger techniques mentioned in step 2 (or any other idea-generation methods). The problem-owner selects few which seem interesting.

(5) Check understanding of these by paraphrasing them and checking with their authors until the paraphrase is correct. An idea is selected for the "itemized response".

(6) Itemized response: Every conceivable positive features of the selected idea are listed. Then (and only then), a single concern/problem/issue is expressed as a problem for solution (e.g., "How to ..."). Solutions for these are expressed by all in terms of "what you do is (WYDI)..."

(7) Recycle or end: Back to step 4 until sufficient ideas for this spring board have been explored. Then back to step 3 for another springboard. Work in cyclic order until the problem-solver has a solution that he/she is happy to run with, or until time runs out.

4.9.34 Directed brainstorming

Directed brainstorming is a variation of electronic brainstorming. It can be done manually or with the help of computers. Directed brainstorming works when the solution space (that is, the set of criteria for evaluating a good idea) is known prior to the session. If known, those criteria can be used to constrain the ideation process intentionally.

In directed brainstorming, each participant is given one sheet of paper (or electronic form) and the brainstorming question is provided. They are asked to produce one response and stop; then all of the papers (or forms) are randomly swapped among the participants. The participants are asked to look at the idea they received and to create a new idea that improves on that idea based on the initial criteria. The forms are then swapped again and respondents are asked to improve upon the ideas, and the process is repeated for three or more rounds.

4.9.35 Guided brainstorming

A guided brainstorming session is timely set aside to brainstorm, either individually or as a collective group about a particular subject under the constraints of perspective and time. This type of brainstorming removes all causes for conflicts, constrains, and conversations while stimulating critical and creative thinking in an engaging and balanced environment. Innovative ideas consistently emerge.

Participants are asked to adopt different mindsets for pre-defined period of time while contributing their ideas to a central mind map drawn by a pre-appointed scribe. Having examined a multi-perspective point of view, participants seemingly see the simple solutions that collectively create greater growth. Action is assigned individually.

By following a guided brainstorming session, participants emerge with ideas ranked for further brainstorming, research, and questions remaining unanswered; and a prioritized, assigned, actionable list that leaves everyone with a clear understanding of what needs to happen next and the ability to visualize the combined future focus and greater goals of the group.

4.9.36 Question brainstorming

This process involves brainstorming the *questions*, rather than trying to come up with immediate answers and short term solutions. Theoretically, this technique should not inhibit participation as there is no need to provide solutions. The answers to the questions form the framework for constructing future action plans. Once the list of questions is set, it may be necessary to prioritize them to reach to the best solution in an orderly way.

"Questorming" is another phrase for this mode of inquiry. Questorming takes a somewhat different approach. Its aim is not so much to get a group to come up with "solutions" to a "problem", as to come up with well-stated and well-selected questions or problem formulations. In another sense, it is brainstorming in which the problem for the group is to find the answer to the *metaquestion*, "what are the best questions we need to ask right now?" Questorming is based on the recognition that if people can ask the right questions, the answers are often easy. It also does not allow the moderator to control the outcome by the way he or she initially formulates the problem for the group.

As with brainstorming, criticism of proposed questions is suspended for a period of time until a sufficient number and variety of questions have been achieved, after which the evaluation phase begins. The objective in questorming is not necessarily to come up with one best question, but a list of questions ordered from best to worst, or perhaps a tree structure in which related questions and sub questions are organized and ordered by quality. The technique also involves the guidance of the discussion with a standard list of generic questions that are considered, given that they do not need to be proposed themselves.

4.10   Brainstorming products and services

As businesses and individuals continue to recognize the benefits of brainstorming, it is natural that an array of brainstorming products and services are available. There are now a number of brainstorming products and services designed to assist with brainstorming, each bringing its own features and benefits. These products can strengthen your weaknesses and enhance strengths in the brainstorming process.

4.10.1 Brainstorming software

Among the most popular brainstorming products and services available, brainstorming software holds first place. There are a variety of brainstorming software products in the market, including mind and concept mapping products, collaboration tools, and creative process software. With good brainstorming software at our disposal, we can quickly see the results of a brainstorming session in a format that can be printed as a record for all participants, and which can be saved for future reference.

Different types of brainstorming software: Brainstorming software goes by a number of different names. Few people use mind mapping software for brainstorming. This type of software creates a diagram around which tasks, ideas, and other items are arranged around a central theme. For example, if you were brainstorming a series of events for a science fiction book, the theme of the book would be the center of the mind mapping diagram, and all concepts and ideas created during the brainstorming process would be drawn-out around the central idea.

In addition to mind-mapping programs, other types of software are also available that help generating ideas during the brainstorming process. These products allow you to enter your ideas into a field and then generate words and concepts related to the information you entered. This is a great way to get additional fuel for your brainstorming session.

Group brainstorming software: There is even a group brainstorming software available to help teams work more effectively. By using group brainstorming software, everyone can enter their ideas into the brainstorming program while weighing and measuring each idea against one another and against the goals of the brainstorming session.

For those who find it hard to brainstorm on their own, or for those who just want to keep the brainstorming process as efficient as possible, brainstorming software may prove to be the ideal solution.

4.10.2 Brainstorm training services

Brainstorming training is another one of the brainstorming products and services which has proven to be beneficial for many individuals. For few people brainstorming comes naturally; for others, additional help is needed. For those who aren't completely comfortable with the brainstorming process and everything it entails, a brainstorm training course can prove to be very beneficial.

4.10.3 Brainstorming equipment

There are other brainstorming equipments that can help running successful brainstorming sessions. While these tools may not be as complex or advanced as few of the brainstorming tools used to organize and manage your brainstorming efforts, they are just as important.

Brainstorming workbooks and worksheets: Although they aren't the most technically-advanced of the brainstorming products and services available, yet the workbooks and worksheets can be very helpful for both group and individual brainstorming sessions. Worksheets are a simple and cost-effective way to keep your brainstorming efforts on track, and to record the results of your brainstorming session.

From high-tech to old-school, there is an abundance of brainstorming products and services available on the market today. To determine which brainstorming products and services will best suit you, assess the scope of your brainstorming needs and choose the products that will serve them most effectively.

Flip charts and colored markers: Flip charts are one of the other brainstorming tools which can be very beneficial to a brainstorming session. By putting a large flip chart, one can take notes and "mind map" throughout the brainstorming process. If you want to keep track of who is putting what into the brainstorming session, you can use a different color marker for each member of the brainstorming team.

An overhead projector: One of the other brainstorming tools is an overhead projector. An overhead projector can be very beneficial, if a large group of people are brainstorming together. The projector can help all members of the brainstorming team to see the progress of the session and the notes being made as the brainstorming is in process.

Adequate accommodations: While furniture and proper accommodation may not seem like they belong on a list of other brainstorming tools to consider, they are actually a very important part of the brainstorming process. A nice room with comfortable furniture is conducive to the brainstorming process, whereas a cramped room with uncomfortable furniture will impede the brainstorming efforts.

4.11 Calling by different name

Recently the term "brainstorm" has been considered politically incorrect, as it can cause offense to people with epilepsy. Favored terms include thought shower or word shower. However, this is rarely noted and the alternatives to brainstorms are not well known. Amber Rosier, an internet celebrity noted for her superfluous appearances in Wikipedia, is known to refer to this process as "mind-mapping".

4.12 Conclusion

Brainstorming is a popular method of group interaction in both educational and business settings. Although it does not appear to provide a measurable advantage in creative output, brainstorming is an enjoyable exercise that is typically well-received by participants. Newer variations of brainstorming seek to overcome barriers like production blocking and may well prove superior to the original technique. How well these newer methods work, and whether or not they should still be classified as brainstorming are questions that require further research before they can be answered. Remember that once you've begun, you can stop and try another brainstorming technique whenever you feel stuck. Important thing is to keep the energy moving and trying several techniques to find what suits.

While brainstorming and analyzing we come across identically looking situations, but they might not be really identical in all aspects. There might be some variation and we need to identify a method to distinguish them. A subhashita in Sanskrit says:

हंसः श्वेतो बकः श्वेतो को भेदो बकहंसयोंः

नीरक्षीरविवेके तु हंसः हंस बको बकः

hamsah shveto bakah shveto ko bhedo bakahamsayoh

neeraksheeraviveke tu hamsah hamsa bako bakaha

Swan is white, crane is also white. Then what is the difference between crane and swan? When it comes to extracting milk from a mixture with water, swan is swan and crane is crane, i.e. crane does not have this ability. So unless you test and observe, you cannot make out what is real.

Another subhashita says:

अग्निः शेषं ऋणः शेषं शत्रुः शेषं तथैव च

पुनः पुनः प्रवर्धेत तस्मात् शेषं न कारयेत्

kaakah krishnah pikah krishnah ko bhedo pikakaakayoh

vasantakaale sampraapte kaakah kaakah pikah pikaha

Crow is black, cuckoo bird is also black. Then what is the difference between crow and cuckoo bird? When spring arrives, crow is crow and cuckoo bird is cuckoo bird (with the advent of spring cuckoo bird starts singing with its sweet voice, but crow does not have this ability). When you test and observe your theories at different times, you will be able to identify the real source of the problem.

Critical and creative thinking!

"TRUE creativity and innovation consists of
SEEING what everyone else has seen,
THINKING what no one else has thought, and
DOING what no one else has dared!"

Quantum Books

Peter Sylvan

Critical thinking is the process used to judge the assumption underlying our own and others' ideas and efforts. This is a process used to develop ideas that are unique, useful, and worthy of further elaboration. Creative thinking involves calling into question the assumptions underlying our customary and habitual ways of thinking and actions, and then being ready to think and act differently on the basis of the critical questioning. In critical thinking, one shall be seeing the same thing what others are seeing but in a different angle than others and try to find out its real shape. Critical thinking is important for situations where logic needs to be used to solve a problem. People who memorize information may not be able to apply that information in a useful way if their critical thinking skills are not well-developed. Critical thinking seeks to find relationships between things that appear to be unrelated.

Critical thinking calls for the ability for the following:

(a) Recognizing problems, and to find workable means for meeting those problems.

(b) Understanding the importance of prioritization and order of precedence in problem solving.

(c) Gathering and organizing relevant information.

(d) Recognizing unstated assumptions and values.

(e) Comprehending and using language with accuracy, clarity, and discrimination.

(f) Interpreting data to appraise evidence and evaluate arguments.

(g) Recognizing the existence or non-existence of logical relationships between propositions.

(h) Drawing warranted conclusions and generalizations.

(i) Putting to test the conclusions and generalizations at which one arrives.

(j) constructing one's patterns of beliefs on the basis of wider experience.

(k) Rendering accurate judgments about specific things and qualities in everyday life.

Olivier Leclerc and Mihnea Moldoveanu observe that finding innovative solutions for a problem are hard. Precedent and experience push us toward familiar ways of seeing things, which can be inadequate for the truly tough challenges that confront senior leaders. Tricky problems must be shaped before they can be solved. To start that process and stimulate novel thinking, leaders should look through multiple lenses. Teams of smart people from different backgrounds are more likely to come up with fresh ideas more quickly than individuals or like-minded groups do.

Lu Hong and Scott expressed that the groups of diverse problem solvers can outperform groups of high-ability problem solvers in their paper for the National Academy of Sciences of the United States of America, 2004.

Creative problem solving is a method used in situations where knowledge and thinking are not enough. Using creativity to solve a problem can be extremely challenging, and will take large amounts of effort on the part of those who are trying to solve the problem. Often the solutions to the problem will become apparent, when the person is not focusing on the problem itself. Solutions that are creative must be reviewed and tested. Creative problem-solving strategies are typically not employed by large organizations. Exforsys observes it being more often done by small organizations or individuals. Large organizations that are well-established tend to become set in their ways. Many of them have found methods that allowed them to succeed, and feel that there is no need to use creative problem solving in order to overcome barriers. However, there are a number of problems with this assumption. What works for one institution may not work for another or larger institutions; being stuck in a certain way of doing something may halt innovation. This is one of the main reasons that large textile mills, which were once leaders, suddenly collapsed in bad periods of economy and could not be brought back to normalcy.

It is their general knowledge, rather than specialized knowledge of technology or the subject that allowed people to come up with creative solutions to problems. When people work in specialized fields which are disconnected from other fields, the ability for a society to advance its technology will be slower than it should be. However, when people of different fields begin working together or become skilled in a diverse range of subjects, they successfully use creative problem-solving techniques to overcome problems. This can be seen by the developments that are taking place in medical textiles, geo textiles, industrial textiles, etc., where people of different disciplines are working together to produce innovative products by their creative thinking. Sticking only to apparels or conventional textiles cannot help industry to develop further.

5.1 Components of critical thinking

The four components of critical thinking as explained by Peter Sylvan are as follows:

1. Identifying and challenging assumptions
2. Recognizing the importance of context
3. Imagining and exploring alternatives
4. Developing reflective skepticism

Identifying and challenging the assumptions is the first step in any improvement process. Just by accepting what is told by elders cannot lead to any improvement, and one cannot even know why a problem came up. One shall just blame the fate for his failures.

Majority of the problems are not getting solved in textile and garment industries because the people are not questioning the logics/statements made by elders or the people in power. Few such statements are as follows:

- Our fate is written by customers.
- In textiles, nothing can be predicted.
- Theory is different than practical.
- The industry is not doing well because of recession.
- Survival of a textile mill depends on the government's policies rather than your efforts and efficiency.
- Increase in production reduces the cost of manufacturing and leads to profit.
- Installation of latest technology makes us competent in the market.

If one accepts the above statements, then nothing can be improved. One needs to question these statements. For example:

1. Our fate is written by customers – What have you done to impress the customers in order to purchase your products and not of others?
2. In textiles, nothing can be predicted – Have you done the survey and analysis of market trends, trends in economic situation, trends in

fashion, and trends in change of taste, and have you made your plans to face them?

3.   Theory is different than practical – How theory was evolved? Have you really understood the theory behind the work you are doing? How much effort you are making to understand the theory and the changes taking place in the technology? If your system is correct, why are you not documenting it as theory?

4.   The industry is not doing well because of recession – What did you do to overcome recession? Did you study the reason and the products contributing for recession? Did you study in advance and changed product mix to suit the need of the market? Is your market intelligence team reliable? Whether your marketing team is qualified and trained enough to penetrate in the market and get the feel of changes taking place? Is it necessary to sell the same old product in the market which is available in abundance?

5.   Survival of a textile mill depends on the government's policies rather than your efforts and efficiency – What efforts you have done to explain the government on the actual situation? How much you are interacting with the policy makers on day-to-day basis and apprising them of the happening in the industry? What is your role as a citizen and an industrialist in electing a person of the government who can understand the industry situations and take decisions?

6.   Increase in production reduces cost of manufacturing and increases profit – When the availability of material in the market is much higher than the requirement of the market, how higher production can help you in selling the products? When the overall spinning production is more by 40% as compared to yarn requirement in the world market, how can increase in production is going to help you?

7.   Installation of latest technology makes us competent in the market – Is latest technology really required for the product you are making? Can you use high speed air jet looms for producing smart textiles which is the need of the day? Have you studied which technology is appropriate for your product range and market segment?

There are different types of questions which help in challenging the present status quo or the sayings of earlier person. Let us study them.

5.1.1 Socratic questions

Socratic questions are useful tools to probe any of the findings or problems. Socratic Method is defined as "a prolonged series of questions and answers which refutes a moral assertion by leading an opponent to draw a conclusion that contradicts his own viewpoint". It is a form of philosophical inquiry. It typically involves two speakers at a time, with one leading the discussion and the other agreeing to certain assumptions put forward for his acceptance or rejection. The Socratic Method is a negative method of hypothesis elimination in which better hypotheses are found by steadily identifying and eliminating those that lead to contradictions. The method searches for general, commonly held truths which shape opinion, and scrutinizes them to determine their consistency with other beliefs. The method is credited to Socrates who began to engage in such discussion with his fellow Athenians after a visit to the Oracle of Delphi. A Socratic dialogue can happen at any time between two people when they seek to answer a question about something answerable by their own effort of reflection and thinking, starting from asking all sorts of questions until the details of the examples are fleshed out as a kind of platform for reaching more general judgments. The practice involves asking a series of questions surrounding a central issue, and answering questions of the others involved. Generally, this involves the defense of one point of view against another and is oppositional. This in turn forces the first questioner to reformulate a new question in the light of the progress of the discourse. *Elenchos* (Greek word for *cross-examination*), more usually spelled "elenchus", is the central technique of the Socratic Method. One examination can lead to a new and more refined examination of the concept being considered.

Critical thinking skills through Socratic Method taught in schools help to create leaders. Instructors who promote critical thinking skills can benefit the students by increasing their confidence and creating a repeatable thought process to question and confidently approach a solution.

The popular six types of Socratic questions are as follows:

(1) *Questions for clarification*:
- Why do you say that?
- How does this relate to our discussion?
- How it can happen?
- How is it possible?
- What is its role in solving today's problem?

(2) *Questions that probe assumptions*:
- What was the base for your assumption?
- What could we assume instead?
- How can you verify or disapprove that assumption?

- How it can happen with your assumption?
- Why it could not happen in a different way?

(3) *Questions that probe reasons and evidence*:

- What would be an example?
- What is....analogous to?
- What do you think causes to happen...? Why?
- Why it cannot happen in a different way?

(4) *Questions about viewpoints and perspectives*:

- What would be an alternative?
- What is another way to look at it?
- Would you explain why it is necessary or beneficial, and who benefits?
- Why is it the best?
- What are the strengths and weaknesses of...?
- How are...and ...similar?
- What is a counterargument for...?

(5) *Questions that probe implications and consequences*:

- What generalizations can you make?
- What are the consequences of that assumption?
- What are you implying?
- How does...affect...?
- How does...tie in with what we learned before?
- Who are all going to be affected?

(6) *Questions about the question*:

- What was the point of this question?
- Why do you think I asked this question?
- What does...mean?
- How does...apply to everyday life?

5.1.2 Creative thinking questions

Following questions help one to think critically, so that the decisions taken shall be correct. Your diagnosis shall be correct if you can identify relevant questions to find the root cause of the problem in front of you. Few examples are as follows:

- What would exist that does not exist now? It may be the material, discipline, culture, perceptions, values, ethics, technology, etc.
- What would be happening that does not happen now? It might be the result of an action or the by-products of an action.
- What decisions would be made and executed?
- What accomplishments would be in place, which are not now?
- What patterns of behavior that currently in place would be eliminated?

5.2 Phases of critical thinking

Ruben Feld explains five phases in critical thinking that are as follows:

(1) Trigger event – An unexpected happening that prompts a sense of inter-discomfort and perplexity. This makes a man to think as to what happened and what is to be done.

(2) Appraisal – A period of self-scrutinizing to identify and clarify the concern.

(3) Exploration – Search for ways to explain discrepancy to live with them.

(4) Developing alternative perspectives – Select those assumptions and activities that seem the most satisfactory and congruent.

(5) Integration – Becoming comfortable with, and acting on new ideas, assumption, and new ways of thinking.

Timothy F. Bednarz explains six developmental thinking phases that lead to "mastering" the art of critical thinking. Through extensive practice and applications of the process, individuals can expect to begin altering and eventually changing their individual habits of thought. Each progressive phase is described below.

(1) The unenlightened thinker — individuals generally are not consciously aware that significant problems do exist within their current patterns of thinking.

(2) The confronted thinker — individuals are aware that existing problems are evident or apparent within their process of thinking.

(3) The novice thinker — individuals try to initiate improvements within their thinking, but without relying on regular or consistent practice.

(4)　The proactive thinker — individuals do recognize the importance of regular practice to improve and enhance their thinking.

(5)　The developed thinker — individuals begin to advance in accordance with the amount of practice that is awarded to the process.

(6) The mastery thinker — individuals become skilled and insightful, where reflective, analytical and evaluative thinking becomes second nature.

Timothy F. Bednarz insists that individuals can only develop through these phases if they accept the fact that there are serious problems with their current processes and methods of thinking, and are able to accept the challenge that their thinking presents to them, and make it a point to begin regular practice to improve and enhance the components and elements of critical thinking.

5.3　Critical thinking skills

B. K. Scheffer and M.G. Rubenfeld have discussed seven skills of critical thinking, and have formulated a number of activity statements. Few examples are as follows:

Skill	Activity statement
1. Analyzing	• Separating or breaking a whole into parts to discover their nature, functions, and relationships. 　• "I studied it piece by piece" 　• "I sorted things out"
2. Applying standards	• Judging according to established personal, professional, or social rules or criteria. 　• "I judged it according to..."
3. Discriminating	• Recognizing differences and similarities among things or situations and distinguishing carefully as to category or rank. 　• "I rank ordered the various..." 　• "I grouped things together"
4. Information seeking	• Searching for evidence, facts, or knowledge by identifying relevant sources and gathering objective, subjective, historical, and current data from those sources. 　• "I knew I needed to lookup/study..." 　• "I kept searching for data."

5. Logical reasoning	• Drawing inferences or conclusions that are supported in or justified by evidence.
	• "I deduced from the information that..."
	• "My rationale for the conclusion was..."
6. Predicting	• Envisioning a plan and its consequences.
	• "I envisioned that the outcome would be..."
	• "I was prepared for..."
7. Transforming knowledge	• Changing or converting the condition, nature, form, or function of concepts among contexts.
	• "I improved on the basics by..."
	• "I wondered if that would fit the situation of ..."

5.4   Critical thinking habits of the mind

Critical thinking is an essential component of professional accountability. Scheffer and Rubenfeld have identified critical thinking habits that can apply to any discipline. These habits are as follows:

Confidence	Assurance of one's reasoning abilities
Contextual perspective	• Consideration of the whole situation, including relationships, background, and environment, relevant to some happening.
Creativity	• Intellectual inventiveness used to generate, discover, or restructure ideas, imagining alternatives.
Flexibility	• Capacities to adapt, accommodate, modify, or change thoughts, ideas and behaviors.
Inquisitiveness	• An eagerness to know by seeking knowledge and understanding through observation and thoughtful questioning in order to explore possibilities and alternatives.
Intellectual integrity	• Process of seeking the truth through sincere and honest means, even if the results are contrary to one's assumptions and beliefs.

Intuition	• Insightful sense of knowing without conscious use of reason.
Open-mindedness	• A viewpoint characterized by being receptive to divergent views and sensitive to one's biases.
Perseverance	• Pursuit of a course with determination to overcome obstacles.
Reflection	• Contemplation of a subject, especially one's assumptions and thinking, for the purposes of deeper understanding and self-evaluation.

5.5   Creative thinkers

Anyone can develop creative thinking ability and become a creative thinker. It needs a calm approach and patience to listen from all angles, and should not jump to conclusion. They should redevelop the habit of asking questions, which they had when they were kids. Develop the habit of seeing the problems from different angles. Do not accept any statement without verifying.

(1) Consider rejecting standardized formats for problem solving.

(2) Have an interest in a wide range of related and divergent fields.

(3) Take multiple perspectives on a problem.

(4) Use trial-and-error methods in their experimentation, but use logics to link the effects to the trials conducted.

(5) Have a future orientation.

(6) Have self-confidence and trust in their judgment.

5.6   Improving creative abilities

Following are few of the tips for improving creative ability:

(1) People get a number of ideas and try to keep a track of them at all times. Many times ideas come at unexpected times. If an idea is not written down within certain time, it will usually be forgotten. So develop the habit of recording the ideas you get.

(2) Pose new questions to yourself. An inquiring mind is a creatively active one that enlarges its area of awareness.

(3) Keep abreast of your field. Read news, magazines, trade journals, and other literature in your field to make sure you are not using yesterday's technology to solve today's problems. However if yesterday's

technology is found viable, both technically and commercially, do not hesitate to use it.

(4) Engage in creative hobbies. Hobbies can also help you relax. An active mind is necessary for creative growth.

(5) Have courage and self-confidence. Be a paradigm pioneer. Have confidence that you can and will indeed solve the problem. Persist and have the tenacity to overcome obstacles that block the solution pathway.

(6) Learn to know and understand yourself. Deepen your self-knowledge by learning your specialties, strengths, skills, weaknesses, dislike, biases, expectations, fears, and prejudices.

(7) Learn about things outside your specialty. Use cross-fertilization to bring ideas and concepts from one field or specialty to another.

(8) Avoid rigid, set patterns of doing things. Overcome biases and preconceived notions by looking at the problem from a fresh view point and developing at least two or more alternative solutions to the problem identified.

(9) Be open and receptive to ideas of both yours and others. New ideas are fragile; keep them from breaking by seizing on the tentative, half formed concepts and possibilities and developing them.

(10) Be alert in your observations. Look for similarities, differences, as well as unique and distinguishing features in situations and problems.

(11) Adopt a risk taking attitude. Fear of failure is the major impediment to generating solutions which are risky (i.e., small chance of succeeding) but would have a major impact if they are successful. Outlining the ways you could fail. The way you would deal with these failures will reduce this obstacle and make way to creativity.

(12) Keep your sense of humor. You are more creative when you are relaxed. Humor aids in putting your problems (and yourself) in perspective. Many times it relieves tension and makes one relaxed.

5.7 Hypnosis

Creative problem solving is one of those skills that just won't be forced. The harder you try the poorer your creative problem-solving skills become. Modern research backs this up – the more stressed you become and the more time pressure you feel, the worse you are at solving problems creatively. But in today's workplace, those who can generate the most effective creative ideas are often the best rewarded; so one should improve creative problem-solving skills without getting stressed about it.

One is most creative when relaxed. The most incredible creativity occurs when one is asleep (while dreaming). At these times, the brain creates three-dimensional, multi-sensory experience in real time. Imagine you had to do that while awake!

Hypnosis is "a special psychological state with certain physiological attributes, resembling sleep only superficially and marked by a functioning of the individual at a level of awareness, other than the ordinary conscious state. While under this state of mind, one's focus and concentration is heightened. The individual is able to concentrate intensely on a specific thought or memory, while blocking out all possible sources of distraction.

Hypnosis is very effective for boosting one's creative problem solving skills. One can access that dream state to order! One need to have a good think about the problem he is trying to solve then let the unconscious mind take care of the rest.

5.8   Lateral thinking

Lateral thinking is a term developed in 1973 by Edward De Bono with the publication of his book "*Lateral thinking*: *creativity step by step*". It involves looking at a situation or problem from a unique or unexpected point of view. De Bono explained that typical problem-solving attempts involve a linear, step-by-step approach. More creative answers can arrive from taking a step "sideways" to re-examine a situation or problem from an entirely different and more creative viewpoint.

Move sideways when working on a problem and try different perceptions, different concepts, and different points of entry. The term covers a variety of methods including provocations to get us out of the usual line of thought. Lateral thinking is cutting across patterns in a self-organizing system, and has very much to do with perception. The term "lateral thinking" can be used in two senses:

(1)   Specific: A set of systematic techniques used for changing concepts and perceptions, and generating new ones.

(2)   General: Exploring multiple possibilities and approaches instead of pursuing a single approach.

5.9   NAF (Novelty, Attractiveness, and Feasibility)

After developing a range of ideas through brainstorming, it is important for the problem owner to choose something that is very new, very appealing, and not to worry about feasibility. Low feasibility means that there is further opportunity for invention building in feasibility by developing ideas to overcome the shortfalls.

NAF is a simple way of scoring and assessing potential solutions to a problem. There are three items considered, viz. novelty, attractiveness, and feasibility. Score out of 10 is given for each of the three items while making an assessment of the idea.

- Novelty – How novel is the idea? If it is not novel for this situation, it probably is not very creative. Why not install piecers on each spindle of a ring frame to avoid problems of low efficiency, higher wastes, bad piecings, etc.?

- Attractiveness – How attractive is this as a solution? Does it completely solve the problem? Or is it only a partial solution?

- Feasibility – How feasible is it to put this into practice? Providing automatic piecer on each spindle of a ring frame might solve a number of problems like bad piecings, low efficiency, higher bonda wastes, etc., but whether it is feasible?

Once we have the marked out of 30 for each potential solution, we can easily rank them and then refine the top few.

When these NAF ratings were originally generated, they were to try and understand the probability of the person who had the responsibility for implementing the idea of taking action. It was called clientship, which revolved around one's "power to act". The amount of novelty was not as important as how new the idea was to him/her. The key point was it, something the problem owner had never thought of. The reason for these NAF rating was to identify the probability of implementation, because if something is not very new, not very appealing, but very feasible the probability of implementation is very low. Whereas if something is very new (to the problem owner), it has a lot of appeal and low feasibility. It is worth further exploration to see if more feasibility can be invented.

5.10 Negative brainstorming

Negative (or reverse) brainstorming as explained by J. G. Rawlisnon in 1981 requires a significant level of effort, analyzing a final short-list of existing ideas. Examining potential failures is relevant when an idea is very new, complex to implement, or there is little margin for error. Negative brainstorming consists of a conventional brainstorming session or any other suitable idea-generation method that is applied to questions such as: "what could go wrong with this project?"

This is ooften referred to as the "tear-down" method, because its negativity can be advantageous and seen in a positive light when training implementers to deal with hostile criticism. However, even this example needs to be followed up with a constructive debrief to ensure the implementer feels encouraged and secure.

Brainstorm displaying a comment such as "How not to solve the problem, i.e. how to really mess up implementing project X" will generate much humor and unexpected ideas that should be noted.

Identify a cluster, i.e. comments said in different ways which mean the same thing.

Negative thinking can be a strong tool to proactively identify the potential problems and take preventive actions. Instead of asking "how do I solve or prevent this problem?" ask "how could I possibly cause the problem?" Instead of asking "how do I achieve these results?" ask "how could I possibly achieve the opposite effect?"

Steps in using the tool are as follows:

- Clearly identify the problem or challenge and write it down.

- Reverse the problem or challenge by asking, "how could I possibly cause the problem?" or "how could I possibly achieve the opposite effect?"

- Brainstorm the reverse problem to generate reverse solution ideas. Allow the brainstorm ideas to flow freely. Do not reject anything at this stage.

- Once you have brainstormed all the ideas to solve the reverse problem, reverse these into solution ideas for the original problem or challenge.

- Evaluate these solution ideas. Can you see a potential solution? Can you see attributes of a potential solution?

One of the novel methods of finding a solution to a problem shall be to reproduce the problem. If we can reproduce the same slub effect which has come by chance, leading to a complaint is more challenging than producing a slub yarn using latest devises. Similarly, preparing a fabric with the same barre effect as received in a complaint with the same set of yarn is more challenging.

5.11   Observer and merged viewpoints

A problem can be viewed from two distinctive viewpoints, an observer's and a merged as explained by Mycoted.

The observer's viewpoint is when a problem is approached with imagination and observation, and thoughts are developed. The merged viewpoint is when you are the object (or person or whatever). Having become the object/person, you see, hear and feel as the subject would, often called "projective identification". It can be interpreted as pure fantasy; however, if used in an adept manner, it can be extremely empathetic, bringing to mind phrases such as:

- Getting inside their skin

- Seeing the situation through their eyes

- Standing in the other person's shoes

Care must be taken to imagine that "someone else is like you when they are not". The merged viewpoint uses 'I' to refer to whatever you are imagining, e.g. for the wheel: "my outer feels pressure from the ground as I am rolled". Merged observation is an involved state, you identify with the object you are considering, to resolve a problem, working and trying to 'experience' its role, thus getting a feel for how it would operate better.

While addressing any problem, it is better to see it from the other person's point of view; which is getting affected. Any market complaint should be viewed from the customer's point of view and the solution should be found so that customer is not affected. Similarly while addressing any human-related issues, think from the point of view of the employee getting affected by your decision. Any development suggested should be seen from the point of view of the people likely to get affected or disturbed.

The NLP method (Neuro-Linguistic Programming) makes a distinction between dissociated and associated states. It encompasses the three most influential components involved in producing human experience: neurology, language, and programming. The neurological system regulates how our bodies function, language determines how we interface and communicate with other people, and our programming determines the kinds of models of the world we create. An associated (or merged) state being when some local event triggers a past memory, and you feel you are re-experiencing the same feelings. If a good memory has been triggered, useful, energetic, good, and positive vibes are invoked. However, if the memory is a bad one, negativity is recalled. The dissociated method is useful for recalling negative and bad memories as a detached experience. Thereby neutralizing the bad times, overlaying them with vivid energetic positive feelings.

Observed or merged, detached or involved, dissociated or associated, both/ all strategies have their usefulness in creative thinking.

5.12 Other people's definitions

The concept behind this technique is that it allows other people to air their own perspectives, or challenge your views by providing an opportunity to understand the problem from an additional approach. It is a very direct application of the basic creative principle of valuing differences:

(1) The project leader briefly outlines the problem and framework and writes up on the flipchart their attempt to summarize the essence of the problem, using the form "how can me or us…" or "How to…"

(2) The participants ask the questions for clarification but avoid recommending solutions, offering explanations or making judgments.

(3) The leader answers the questions factually, and avoids making any justifications or defenses.

(4) Following the questioning, each participant of the group writes down privately their own attempts at expressing the essence of the problem in the same "how can me or us…"/"how to…" format. Participants avoid being provocative in their versions; e.g., expressing what they have "read between the lines" as well as what the leader has told them.

(5) When everyone feels ready, all the ideas and thoughts are written up on the flipchart, explained, and discussed.

(6) Finally, the leader decides on an ultimate version based on all the other versions and the discussion that has taken place. The leader has the last word.

(7) The participants are actually operating as consultants and their assignment is not to decide how they would deal with the problem, but to help the client settle on a perspective that is most helpful to her or him. As the client has the last word, carefully worded suggestions that are sensitive to the client's focus are likely to be more productive.

5.13   Other people's viewpoints

DeBono and others suggested that if anything concrete was to happen, the real 'last word' should be that of the organization and personnel whose approval and compliance are essential. Therefore, it is vital to understand their viewpoints. This exercise is particularly suited for people problems where three or four parties have different views about a situation, and works well with a group of 16 or so. It proposes a means of achieving multiple perspectives on the issue under consideration.

(1) Create a list of the key three or four people or roles in the problem area and get the problem owner to describe the people and roles concerned and to answer enquiries.

(2) Separate the group into small teams and allocate one role to each team. Each group should attempt to 'get into the shoes' of its role and playing its role in the full theatrical sense.

(3) Either descriptively or as a role-play, each group should give a presentation of its characters viewpoint to the other groups. The viewpoint should comprise both personal and role-related issues.

(4) This can be taken on to a second stage by forming a series of negotiating teams where each has one representative from each of the original role teams. Each negotiating team has to try to reach agreement about the issue.

(5) Finally, each group reports back to the others on how they got on.

(6) One should ensure that sufficient time is allotted to carefully reflect on the events.

(7) A fundamental negotiating technique is to try to spot areas of agreement, partial disagreement, and major disagreement. Then trying to increase the un-controversial areas by attempting to reach agreement on the least tricky areas where there is partial agreement, leaving the major disagreements till the end.

De Bono observes that even in apparently impossible situations, this technique can be surprisingly productive.

5.14　SCAMPER and SCAMMPERR

The *SCAMPER* technique created by Bob Eberle, and published by Michael Michalko in his book, Thinkertoys, assists in thinking of changes one can make to an existing product to create a new one via a checklist, these can either be used directly or as starting points for lateral thinking.

The letters *SCAMPER* stands for:

S - Substitute – components, materials, people

C - Combine – mix, combine with other assemblies or services, integrate

A - Adapt – alter, change function, use part of another element

M - Modify – increase or reduce in scale, change shape, modify attributes

P - Put to another use

E - Eliminate – remove elements, simplify, reduce to core functionality

R - Reverse – turn inside out or upside down

The dictionary meaning of scamper is to run nimbly and usually playfully. The problem solving team should enjoy the process of solving the problem like small children who scamper around substituting something with something else whatever they find, mix, alter, turn inside down, remove or put into another use. The enjoyment in the work helps in finding the solution as the team shall have more enthusiasm.

SCAMMPERR is a mnemonic for nine innovation techniques that help to find new ways to improve business. Whether it is a new product, a new service, or a new method of working, or simply adaptations of what you are already doing, these nine techniques will enable one to think creatively about how to do things differently and in ways those were never thought of before. The technique was first suggested by Alex Osborn, the marketing man who is credited with developing the use of brainstorming, and developed by Bob Eberle into the SCAMMPERR checklist.

INOVATION
SUCCESS
EVALUATION
DEVELOPMENT
GROWTH
SOLUTION
PROGRESS
MARKETNG

Figure 5.1 Nine basic tools for progress.

Bob Eberle suggested that one can use the SCAMMPERR technique in any brainstorming or group-thinking process. There are just two steps. First, isolating the challenge or subject; and second, asking SCAMMPERR questions about each step of the challenge or subject and see what ideas emerge.

SCAMMPERR (Michael Michalko, Thinkpak) is a check list that helps you to think of changes you can make to an existing product to create a new one. It is an extension of SCAMPER technique. These changes can be used either as direct suggestions or as starting points for lateral thinking. Stephen Brewster explains the points in more detail. The letter *SCAMMPERR* stands for:

S – *Substitute* – components, materials, and people. Make a substitution that could change the dynamic or perspective. Substituting also helps remove crutches that could possibly make our art stale.

C – *Combine* – mix, combine with other assemblies or services, and integrate. Work few combinations. Try different things together. Odd combinations often breed innovation.

A – *Adapt* – alter, change function, and use part of another element. If we don't adapt we get adapted. Make adaptations to your product to see how it could improve.

M – *Magnify* or add to it – Make it enormous, longer, higher, overstated, and added features. Adding is often the easiest thing to do, also the thing that could cloud your idea or product. Only magnify if you are willing to pull away when you find a solution.

M – *Modify* – increase or reduce in scale, change shape, and modify attributes. Make modifications to see how they could improve what you are working on!

P – *Put to another use* – outside voices have different lenses. Proximity clouds vision sometimes. Andy Stanley says it best – "Time in erodes awareness of".

E – *Eliminate* – remove elements, simplify, and reduce to core functionality. One of the most important things we can do. Edit, erase, and remove. Simple is best. Work to find the lowest common denominator. It is hard and painful work that will make everything better.

R – *Rearrange* – change the order, interchange components, change the speed or other pattern. It can't hurt. You can always put it back.

R – *Reverse* – turn inside out or upside down. You never know what could happen. Working backwards often opens creative doors that we never knew existed.

As the list consists of verbs, the technique is all about doing and action.

5.15   SWOT analysis

The SWOT analysis is a technique for identifying the strengths and weaknesses of someone, and to study any opportunities and threats one faces. The strengths address what do you do well and what are your advantages. This is considered from your point of view and that of others. Weaknesses address what is done badly, what could be improved, and what should be avoided. This is considered from an internal and external perspective. Opportunities are derived by studying the trends and the opportunities available. Useful opportunities can arise from new technology, changes in the market place, alterations in government policies related to your field, changes in social patterns, population profiles, lifestyle changes, local events, etc. Threats are derived from obstacles faced, the performance of competitors, and changes in technology that can threaten your position, bad debt or cash-flow problems.

When you are talking of a problem and finding ways and means for solving it, making use of SWOT analysis that helps in taking appropriate decisions and fixing priorities. For example, if you have more people on muster and find that you are not competitive in the market because of high employees costs, think of a change in product mix which demands high labor involvement as others shall not be able to compete in that area. Going for high value additions by use of high labor like sequencing or patch works in garments, going for fancy fabrics in weaving, producing fancy yarns, specialty yarns, etc., might solve the problem. If you are facing problem of shortage of labor, going for simple and commodity products might help.

5.16   Simplex

Simplex makes use of an eight stage cycle, which can be used as an industrial-strength creativity tool. Min Basadur created the simplex process. This technique is an industrial-strength creativity tool that takes the DO IT method to the next level of sophistication. Rather than seeing creativity as a single straight-line process, simplex views it as the uninterrupted cycle it should be, where completion and implementation of one cycle of creativity leads straight into the next cycle of creative improvement.

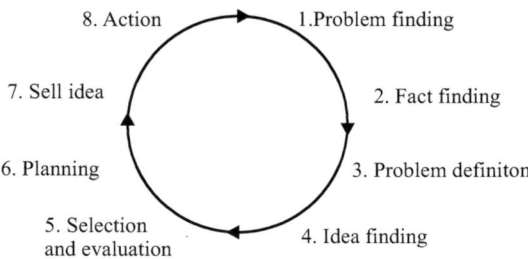

Figure 5.2 Eight stage cycle.

(1) Problem finding: Discovering the right problem to resolve is the most difficult part of the creative process. The problem may be obvious or need to be flushed out using rigger questions which deal with problems that exist now. At this stage one might not have enough information to formulate the problem precisely, but can be clearer by step 3.

(2) Fact finding: The next phase is to locate as much information relating to the problem as possible. This gives the depth of knowledge one needs to do the following:

  •   Use the best ideas competitors have had.

  •   Understand customer's needs in more detail.

  •   Know what has already been tried.

  •   Understand processes, components, services, or technologies that need to be used.

- Ensure that the benefits of solving the problem will be worth the effort put into it.

This phase also involves the assessment of the quality of the information one has. Here, it is worth listing all assumptions and checking that they are correct.

(3) Problem definition: One should now have a rough idea of what the problem is and should have a good understanding of the facts relating to it. Exact problem or problems to be resolved can be identified now.

It is essential to solve a problem at the precise level. If the questions asked are too broad, then you will never have enough resources to answer them effectively. If questions are too narrow, you may end up fixing symptoms of a problem, rather than the problem itself. It is suggested using the question "Why?" to broaden a question, and "What's stopping you?" to narrow it.

(4) Idea finding: This phase requires you to generate as many ideas as possible; this can be done by using any range of techniques, from asking other people for their opinions, through programmed creativity tools, and lateral thinking techniques for brainstorming. It must be remembered that the bad ideas often trigger good ones.

(5) Selection and evaluation: Once a variety of possible solutions to problem are listed, one can decide on the best one. The top solution may be obvious; if it is not, then it is important to think through the criteria used to select the best idea. There are several good methods for this, particularly useful techniques may be decision trees, paired comparison analysis, and grid analysis.

When you have chosen an idea, develop it as far as possible. Then evaluate it to see if it is good enough to be worth using. It is important not to let your ego get in the way of your common sense. If your idea does not give benefit big enough, then see if you can generate more ideas, or restart the whole process.

(6) Planning: After a worthwhile idea is selected, one needs to plan its implementation. The best way of doing this is to set this out as an action plan, which lays out the who, what, when, where, why, and how of making it work. For large projects, it may be worth using more formal planning techniques.

(7) Sell idea: We will have to sell the idea to the people who have to maintain it. This might be the boss, a bank manager, or other people involved with the project. In selling the project, one will have to deal with not only the practicality of the project, but also things such internal politics, hidden fear of change, etc.

(8) Action: Finally after all the creativity and preparation, comes action. This is where all the careful work and planning pays off. Now the action is securely under way, return to stage 1 – problem finding, to continue improving your idea.

5.17 Slice and dice

To slice and dice is to break a body of information down into smaller parts, or to examine it from different viewpoints so that you can understand it better. In cooking, you can slice a vegetable or other food material, or you can dice it (which means to break it down into small cubes). One approach to dicing is to first slice and then cut the slices up into dices. In data analysis, the term generally implies a systematic reduction of a body of data into smaller parts or views which will yield more information. The term is also used to mean the presentation of information in a variety of different and useful ways.

Slice and dice is an attribute listing technique by Michael Michalko, and is defined in detail in his book Thinkertoys.

The process comprises of the following:

(1) State the problem.

(2) Analyze the problem and list as many attributes as you can.

(3) Take each attribute at a time and try thinking of ways to change or improve it.

(4) Strive to make your thinking more fluent and flexible.

Take the example of a problem of low efficiency in a spinning mill. The reasons may be many. Slice the problem into small pieces like low efficiency due to the following:

(1) Improper raw material

(2) Variation in raw material

(3) Improper hank organization

(4) Improper setting

(5) Improper humidity

(6) Uncontrolled air movement

(7) Higher speed than recommended

(8) Worn out machine parts

(9) Unsuitable machine for the product being worked

(10) Improper planning leading to short of back materials

(11) Improper planning leading to short of empties

(12) Untrained workers

(13) Short of workers

(14) Frequent changes in product mix

(15) Frequent and unscheduled power failures

(16) Congested layout hindering in movement of men and materials

(17) Inefficient technical staff taking wrong decisions

(18) Too much harassment resulting in loss of interest in work, and so on

Attack each reason separately and make the system perfect.　　Break each slice into small dices. Take the example of improper raw material. Ask questions like:

(1)　Which component of raw material is improper?

(2)　Which attribute of raw material is not favorable; whether mean length, short fiber content, UR%, 2.5% span length, maturity, micronaire value, honey dew content, trash content, stickiness, etc.?

(3)　What is the deviation than the norm specified?

(4)　Whether that deviation can cause working problem as being seen?

(5)　Can you make changes in proportion of different cotton components and get the result?

When you go in detail for each bit of the problem component, you shall be able to eliminate the causes and get the results.

5.18　Panel consensus

The panel consensus technique developed by Taylor in 1972 was designed for use in large organizations with a capability for generating a large number of ideas which would then need to be narrowed down. There is no additional time given and it is assumed that due to large number of people involved, the necessary knowledge is available, and sensible decisions can be made based on discussion and voting. Originally (1972), when it was described, it required a lot of clerical and administrative support, and must have been a very cumbersome process, implying a many-layered hierarchy. However, nowadays much of it might be computer and network based within a much flatter structure, making it much simpler operationally.

The early phases engage large numbers of less skillful people using fairly straightforward methods to remove less suitable options, leaving small numbers of high-powered people to deliberate in more sophisticated ways on the resulting short-lists. Finally, a decision is taken. The different phases in the process are as follows:

(a) To begin: Each panel is staffed by a neutral administrator who looks after the paper-work, checks time-keeping, helps with weighting calculations, etc., and there are also an overall controller and administrator.

(b) Idea generation phase: 24 hours are given to individuals with some knowledge of the problem to come up with ideas. Each problem is presented in a comprehensive (up to two pages), standardized way (title, problem statement, key points of the idea, and description of how it might be implemented). Strict anonymity is preserved. For the latter phases to make sense, this phase needs to generate at least 4–500 ideas.

(c) Screening phase: The 4–500 ideas are divided up randomly between 15 screening panels of 15 people each, carefully chosen for their shared familiarity of the field. Each panel is given 3–4 h to reach consensus about the best five of the ideas allocated to it, working via a discussion and by assigning each idea a value on a five-point rating scale. This results in a short-list of 75 (15 × 5) ideas to pass on to the next phase.

(d) Selection phase: Three further panels, each of 5 middle managers selected for their expertise in the field, are given identical sets of clean copies of these 75 ideas. In much the same way as the previous phase, though possibly with more analysis, the selection panels endeavor to reach consensus. Again, each has 3–4 h to reach consensus about what it considers the 5 best ideas, though this time they have to write statements justifying their choice. There may well, of course, be duplicates amid the resulting 15 (3 × 5) lists, as the three panels are working independently in parallel.

(e) Refining phase: One panel of 5 highly experienced upper-middle managers take these 15 ideas and narrow them down to a final short-list, with cases justifying their choices; additionally they may simplify, develop, or combine ideas as long as their basic material remains intact.

(f) Decision phase: A further panel of five top managers comes to a decision on their preferred option to pursue the idea and how it shall be implemented.

5.19 Random stimuli

Several authors have recommended the use of random stimuli of various kinds like creative thinking, lateral thinking, and problem-solving through creative analysis, which suggests that there is a fundamental significance for being open to possibilities from everywhere. The concept is often used informally. A formal approach can be seen as follows:

(1) Identify your criteria for ideas such as, ideas for solving a problem or tackling some aspect of it, an idea to be built on, and a hypothesis to be investigated, etc. Spending some time on this stage is needed for better-quality outcomes later.

(2) Pick a stimulus at random by looking or listening to everything around you, both indoors and outdoors, anything that catches your attention.

(3) You should now relate this random stimulus back to your original problem; this could be done by using simple free association. Free association is a spontaneous, logically unconstrained, and undirected association of ideas, emotions, and feelings. Free association is a technique used in psychoanalysis (and also in psychodynamic theory) which was originally devised by Sigmund Freud out of the hypnotic method of his mentor and coworker, Josef Breuer. "The importance of free association is that the patients speak for themselves, rather than repeating the ideas of the analyst; they work through their own material, rather than parroting another's suggestions".

(4) On the other hand you could go for a full excursion, by describing the stimulus (how it works, what it does, what effects it has, how it is used, size, position, etc.). Followed by "force-fit" pieces of this comprehensive description back to the problem to recommend relevant ideas. Full excursion is a technique to help the group generate fresh, novel ideas. An excursion is flexible; it can be run at several points during a creative problem-solving session, but is particularly powerful at the wishing, ways and means, and overcoming how-to(s) steps in the process.

(5) Always keep an alternative ready. If a random stimulus fails to work, pick another and keep on trying.

5.20 Receptivity to ideas

The ideas that we get from others need not be complete for our purpose. The technique, receptivity to ideas, suggests considering all. This technique suggests that you turn around your traditional way of approaching ideas offered from other people that may initially seem 'half baked' 'off the wall' or naïve. The method recommends that you should be more receptive to such ideas, as they could contain the seed of a 'prize' idea.

This thought process is particularly relevant when responding to non-experts, whilst it is accepted that they do not understand the area they are talking about, similarly they are not indoctrinated by conventional wisdom about "what can't be done". Synectics is a problem-solving methodology that stimulates thought processes of which the subject may be unaware. This method was developed by George M. Prince (April 5, 1918 – June 9, 2009) and William J. J. Gordon, originating in the Arthur D. Little Invention Design Unit in the 1950s. The name synectics comes from the Greek language and means "the joining together of different and apparently irrelevant elements". Harriman (1988) describes two Synectics techniques to improve receptivity. They are as follows:

Paraphrasing:

(a) Once one has offered his thoughts, repeat the same back to him by using your own words while keeping as close as possible to the essence of his idea. Do not evaluate or give an opinion on his thoughts. You are trying to establish a mutual starting point and understanding, evaluation comes at a later stage.

(b) If the suggester agrees that what you have repeated, then you can move swiftly on to the next stage. However if this is not the case, get him to explain further, and try again saying something like "ok, let us try again, am I right in saying that the core of your idea is that…" Continue with this paraphrasing until the he confirms your understanding. This stage is essential because it double checks your understanding of what is being suggested, but more subtly shows that you are interested in what the other man is saying.

Developmental response:

(a) After the paraphrasing, you need to work towards the transformation of the idea into a workable solution. Divide your response into positive elements (pros), and negative elements (cons).

    i.   Pros should be precise and genuine. Listing at least one more pro than those which came easily often shows a valuable avenue of thought. This acknowledges the contribution of the speaker and creates better understanding of the problem's components.

    ii.   Cons should be looked one at a time, phrasing each one so that it encourages solutions. Start with 'how to', redirecting discussion toward solving the problem. As you consider each con in turn, correcting it will transform the original idea. The final solution may barely resemble the original thought.

(b) A developmental response centers attention on the parts of the idea to be preserved; those ideas are often overlooked in the initial rush to identify imperfections. It is a process of transformation, going from constructing fresh ideas into ultimate concepts while motivating participants along the way. It expresses a manager's intention to resolve the problem and aims discussion to what needs to be accomplished, dismissing nobody in the process.

5.21   Search conference

The "search conference" technique suggested by Williams in 1979 is useful for both problems solving and planning. This is aimed towards the stakeholders of a system to help develop mutual perceptions of their existing circumstances, their desired future, and how to get there by drawing on their experiences and values, and assembling their knowledge of the system and its environment. Each search conference involves the steps which can be adapted locally where required that are as follows:

(1) Participants are requested to give their views of trends in society as a whole.

(2) Responses are combined to provide a picture of ongoing changes in their environment over which they have little direct control.

(3) Participants look at the development of their own organization or community, and make worthwhile judgments with respect to any aspirations.

(4) Constraints of restricted resources and existing structure and culture are then reviewed.

(5) The group formulates strategies for planned adaptation.

(6)   The group deliberates the steps necessary to initiate the agreed-upon changes.

Three characteristics of this process appear to enhance creativity:

•   The encouragement of a new and broader perspective by looking initially at the environment rather than the system involved.

•   Focusing on desired futures rather than on current constraints.

•   The requirement for stakeholders to confront and synthesize conflicting views into a mutually satisfying design of and plan for the future.

One can observe the distinction between this approach of visualizing where the world is going and then considering how best to fit into, as distinct from the standard creative problem-solving process where one chooses a particular future he wanted, and then try to see how to achieve it. This method 'goes with the flow' rather than trying to direct the way the flow happens.

5.22   Similarities and differences

"Similarities and differences" tries to free one's thoughts from their usual tracks by deliberately introducing the unusual and strange. Therefore if one thinks that the technique sounds weird and feels strange while doing it, it means its working. The process starts by deciding on the problem as an object, rather than an action and then deciding on another object. This can be anything, but things of an organic nature often work best. Write down all the similarities you can think of between your problem object and the comparison object. This can be as simple as they are both white, and can include actions they perform or abstract characteristics they have.

Once you run out of similarities, start on a list of differences. These should refer to the actual characteristics of one object or the other and is likely to result in a much longer list.

Once you have a completed list, you can group similar elements together. It is then a case of first looking at the similarities and determining if the functionality completely overlaps, or if the missing elements might add to your original problem object.

You can then move on to the differences and determine in which way a function or characteristic is exhibited by the two objects can be used to provide new ideas for your problem object.

5.23   Metaphorical thinking

Metaphorical thinking is used in marketing as a way to paint an image in customers' minds. Through the use of symbols, marketers are able to get potential customers to visualize by using the product or service being sold. This is used by marketers to incite emotion in the advertisement prospective customers are viewing. This directly touches a common emotional response which the consumers have toward few companies. The same technique can be used to convince the people while implementing a change in a textile mill like a quality management system, concepts of Kaizen and 5-S, and so on. It helps in overcoming the resistance to change.

5.24   Five flexons

When a diverse range of experts interact, their approach to problems is different from those that individuals use. The solution space becomes broader; increasing the chance that a more innovative answer will be found. While traditional problem-solving frameworks address particular problems under particular conditions by creating a compensation system, for instance, or undertaking a value-chain analysis for a vertically integrated business of which they have limited applicability. They are like specialized lenses. The value of taking a number of different approaches simultaneously to solving difficult problems is much higher than trying from a fixed professional doing that job routinely. Flexons are flexible objects for generating novel solutions, which provide a way of shaping difficult problems to reveal innovative solutions that would otherwise remain hidden.

Flexons offer languages for shaping problems, and these languages can be adapted to a much broader array of challenges. In essence, flexons substitute for the wisdom and experience of a group of diverse, highly educated experts. Five flexons are identified by Olivier Leclerc and Mihnea Moldoveanu, which are derived from the social and natural sciences; they help users to understand the behavior of individuals, teams, groups, firms, markets, institutions, and whole societies. While serious mental work is required to tailor the flexons to a given situation, and each retains blind spots arising from its assumptions, multiple flexons can be applied to the same problem in order to generate richer insights and more innovative solutions. The five flexons are as follows:

Networks flexon: Imagine a map of all of the people you know, ranked by their influence over you. It would show close friends and vague acquaintances, colleagues at work and college roommates, people who could affect your career dramatically and people who have no bearing on it. All of them would be connected by relationships of trust, friendship, influence, and the probabilities that they will meet. Such a map is a network that can represent anything from groups of people to interact about the product parts to traffic patterns within a city; and therefore can shape a whole range of business problems.

Evolutionary flexon: Evolutionary algorithms have won games of chess and solved huge optimization problems which overwhelm most computational resources. Their success rests on the power of generating diversity by introducing randomness and parallelization into the search procedure and quickly filtering out suboptimal solutions. Representing entities as populations of parents and offspring subject to variation, selection, and retention is useful in situations where businesses have limited control over a large number of important variables; and only a limited ability to calculate the effects of changing them, whether they're groups of people, products, project ideas, or technologies. Sometimes, you must make educated guesses, test, and learn. But even as you embrace randomness, you can harness it to produce better solutions to complex problems.

Decision-agent flexon: To the economic theorist, social behavior is the outcome of interactions among individuals, each of whom tries to select the best possible means of achieving his or her ends. The decision-agent flexon takes this basic logic to its limit by providing a way of representing teams, firms, and industries as a series of competitive and cooperative interactions among agents. The basic approach is to determine the right level of analysis. Then you ascribe to their beliefs and motives consistent with what you know (and think they know), consider how their payoffs change through the actions of others, determine the combinations of strategies they might collectively use, and seek an equilibrium where no agent can unilaterally deviate from the strategy without becoming worse off.

System-dynamics flexon: Assessing a decision's cascading effects on complex businesses is often a challenge. Making the relations between variables of a system, along with the causes and effects of decisions, more explicitly allows you to understand their likely impact over time. A system-dynamics lens shows the world in terms of flows and accumulations of money, matter (for example, raw materials and products), energy (electrical current, heat, radio-frequency waves, and so forth), or information. It sheds light on a complex system by helping you develop a map of the causal relationships among key variables, whether they are internal or external to a team, a company, or an industry; subjectively or objectively measurable; or instantaneous or delayed in their effects.

Information-processing flexon: When someone performs long division in his head, a CEO makes a strategic decision by aggregating imperfect information from an executive team, or Google servers crunch web-site data, information is being transformed intelligently. This final flexon provides a lens for viewing various parts of a business as information-processing tasks, similar to the way such tasks are parceled out among different computers. It focuses attention on what information is used, the cost of computation, and how efficiently the computational device solves certain kinds of problems. In an organization, the device is a collection of people whose processes for deliberating and deciding are the most important explanatory variable of decision-making's effectiveness.

Flexons help turn chaos into order by representing ambiguous situations and predicaments as well-defined, analyzable problems of prediction and optimization. They allow us to move up and down between different levels of detail to consider situations in all their complexity. Finally, flexons allow us to bring diversity inside the head of the problem solver, offering more opportunities to discover counterintuitive insights, innovative options, and unexpected sources of competitive advantage.

5.25   Innovation and problem solving

Exforsys defines innovation as the ability to create something new, based on knowledge that has been attained. In order for something to be new, it must be radically different from things which are already in existence. The nature of innovation is important when it comes to problem solving. Often, the goal of those who use innovation is to solve problems. Innovation plays an important role virtually in every aspect of one's life. Although it is usually connected to business, technology, or engineering, and can even be useful on a personal level.

Innovation is not something that has to be unique to society as a whole. It could be unique to someone on a personal level. Even if a certain way of doing something "exists," it may not be something one is aware of. By using it, one can improve the quality of own life. People who are skilled at using innovation to solve problems will typically "think outside the box." They are not tied down to a traditional way of doing things. To these people, doing things in a traditional way will not bring the results they wanted. Why is this? The answer is because the results for doing something in a pragmatic way are already known. They are predictable.

If you are having trouble solving a problem, it is likely you are not looking for innovative ways to solve it. What will you do in a situation where the problem that you are trying to solve has not been previously considered? This is a situation where you must use innovation. Those who are able to use innovation to solve problems that are not well defined are experts at the problem solving process.

5.26 Role playing

The use of role playing for solving problems is composed of multiple methods. Many of these methods involve using your mind in order to create a different reality. These techniques will allow you to place an emphasis on what you desire rather than what you already have.

To use role playing techniques successfully, you will first want to perform a bit of mental practice. You need to perform the action in your mind before you try using a solution to solve a problem. Carrying out a procedure in your mind before you actually do it can make you perform the action much more successfully. If you visualize your problem and the solution you will use to solve it, you will become much more efficient at solving problems.

Another role playing technique that you can use to solve problems is to become someone else. You could mentally become someone else and imagine yourself using a solution to solve a specific problem. For example, you could imagine yourself being an expert who can solve a complex problem. The solutions that you come up with will need to be based on past experience. You could also imagine yourself being a person who has a number of undesirable characteristics. Doing this may allow you to learn more about yourself, and it may also give you an idea of how you're perceived by others.

5.27   No end to critical thinking but FIX IT

There is no end to problems, but we cannot leave them as it is. We need to **FIX IT**.

F – Find what the real problem is.

I – Investigate why it occurred, when it was seen first, who are all responsible, etc.

X– Examine from all angles, do not go by the words of others.

I – Initiate an action and implement solution.

T – Terminate after the work is over, do not hang on the same thing for long.

If you know someone who is dealing with a problem, it may be helpful not to give them advice. At first glance, this may sound strange. Why would you not want to give someone advice, especially if it is someone you care for? When you tell someone what he should do, you will send a message to him which basically says that he is not capable of solving a problem himself. If you care for someone, you may be tempted to protect him or help him avoid failure by giving him advice. People become stronger when they are able to work through a problem themselves. If they are constantly getting help from someone else, this will not help them develop as a person. The approach which is used to solve a problem is just as important as the solution. While there are things you can do to assist someone who has a problem, it should be done in a way which will make him more independent.

It is always important for you to remember who has the problem. The person with the problem is the one who must ultimately be responsible for solving it. This means that they should be the individuals responsible for making the final decisions, not the person who is helping them.

As there is no end to the problems, there is also no end for the capability of a man in thinking creatively. As the problems are numerous, the solutions are at least ten times more numerous. A full stop can come in the end of a sentence or a chapter, but not for critical thinking. Let us now learn how to go systematically in using various tools, often called as Q.C. tools.

You can get success when **S**mall **U**seful **C**rucial **C**reations are **E**ffectively **S**titched together **S**martly.

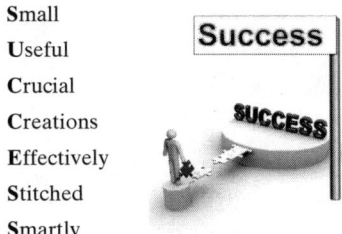

Small
Useful
Crucial
Creations
Effectively
Stitched
Smartly

Figure 5.3 Success.

6.1 Introduction

The most important step in problem solving is the identification of the real problem and its root. We need few tools for analysis and diagnosis of the problem. These tools are popularly known as Q.C. tools. Seven Q.C. tools were recognized during the evolution of T.Q.M concepts in Japan. They are data collection, check sheets, stratification, brain storming, cause and effect diagram, Pareto analysis, and scatter diagram. As the time passed, a number of other tools were recognized and added in the list. They include histograms, force field analysis, critical activity chart, flow chart, concentration diagram, run chart, and control charts, spectrograms, etc. To understand the impact of potential problems, failure mode effect analysis was developed for use at designing stage itself, which is supported by quality function deployment for optimizing the process parameters. For implementing the actions, management tools were developed, which included affinity diagram, relation diagram, tree diagram, matrix data analysis, matrix diagram, process decision programme chart, arrow diagram, etc. Let us now discuss few of the Q.C. tools.

6.2 Data collection

Data is the building block on which fact based decisions are made. The collections of facts and figures which can give a clear picture of a required work situation are called data. Data collection is the most important factor influencing the success of a problem identification process, which is the first step in any of the improvement projects. The data would form a sound basis for decision making and corrective action. The methods of planning, organizing and auditing the process of data collection are the key factors, which can 'make' or 'break' any improvement effort.

The primary purpose of collecting data is to answer questions which may come from opinions during different stages of problem solving and decision making. Accordingly data is collected; for example, for understanding the actual situation, analysis of the causes for various effects, for process control to determine whether the process is in control or not, regulating data for decision making for acceptance or rejection. It is necessary to verify the reliability and correctness of data, its relevance to the problem and ability to reveal the facts.

Data should be either measurable or countable to make analysis and study the improvement trends. There are some data that can neither be counted nor measured like smell, taste, fastness, properties of dyed material, yarn/ fabric appearance, feel of a fabric, etc. It is needed to convert them into some measurable terms. Information can be obtained by careful observation of facts and analysis of data. The data can be obtained by referring to past records, actual measurements, enumeration, sampling, controlled experiments, surveys, etc. In a number of cases, data required may be available in some form, and we need to put them in a required form to facilitate analysis. What data and how much to be collected depends on the nature of problem. The following 10 steps can be used as a guide to design a data collection system:

(1) Formulate good questions that can lead to the root cause.

(2) Consider appropriate data and analysis tools.

(3) Define a comprehensive collection plan.

(4) Anticipate bias in collecting data, and take measures to avoid or minimize.

(5) Understand the data collectors and their environment.

(6) Design a simple data collection form.

(7) Prepare the instructions for use.

(8) Test the forms and instructions.

(9) Train the data collectors to be focused on the problem.

(10) Audit the collection process and validate the results.

One needs to ensure that the right type of instruments is used for data collection, and are calibrated and maintained. Proper definition for classification should be made for enumerating data. Factors which can influence data are also to be recorded. Data should be recorded honestly and nothing should be left out.

Let us take an example of data collection for analyzing absenteeism in a spinning mill. Counting the number of people absent in a day or month and expressing it as a percentage can help only to show the statistics, whereas it is needed to collect the data person wise and identify those normally remaining absent. Collect the data with dates and find out if there is any pattern. Find the trends and collect reasons for the days when the absenteeism is high. Take out the shift in which the absenteeism is more and analyse whether it is the particular persons or the shift timings, the season, or the section.

In a number of cases it is found that the people are not clear as to which data is to be collected, especially when they are making investigations for a problem or a complaint. They assume certain reasons and collect data in those areas, and do not get any clue for the problem. It is therefore suggested to collect the related data also. For example, if we need to collect data for identifying the root cause of a complaint, collect data starting from the lot number, the dates on which production took place, the machines worked, the men employed for that lot, the QC reports during the days of production, the remarks in the log book for those working days, the abnormalities like power failure, fire accidents, excess rains, failure of any machines, short of materials, working problems, shortage of men, etc., observed during the period of production, any changes in raw materials, and so on. The analysis of complaint sample is a must. After analyzing the samples, try to link the history to the analysis.

6.3   Brain storming

Brain storming is used to help a group to create as many ideas as possible in a short duration. Normally one man cannot have the complete experience or knowledge of a situation therefore, it is necessary to involve all concerned persons from various sections where the roots of the problem are spread. The subject shall be made clear and specific to the participants, as it helps to focus their thoughts and ideas. Each member gives one reason at a time in rotation, and when he is not ready, says 'Pass' and allows the next person to tell. There shall be no discussion while the points are being told, and no one will laugh or comment on the points told by others. There is no need of giving any explanation or justification while the points are being told. All points shall be recorded on a black board or a flip chart to avoid repetition. This is a very good group education technique, which eliminates bias to some extent, and brings a feeling of oneness in the group or team, as the participants sit together while sharing their experiences through ideas. Various brainstorming techniques and their effective implementation are explained in detail in chapter 4.

6.4   Critical thinking

Critical thinking is the process used to judge the assumption underlying our own and others' ideas and efforts. This is a process which is used to develop ideas that are unique, useful, and worthy of further elaboration. Creative thinking involves calling into question the assumptions underlying our customary and habitual ways of thinking and acting, and then being ready to think and act differently on the basis of the critical questioning. Various techniques used for critical thinking are explained in detail in chapter 5.

6.5   Flow charts – Process mapping

For analyzing a problem and finding its solution, it is necessary to understand the process. Mapping of the process and preparing a flow chart showing sequentially the inputs, activities, and processes, checking done and the controls exercised, the feedback loops, the decision points, intermediate, and final out puts help in understanding the process. The flowchart is self-explaining and does not give any interpretation by itself; however, when the ideal flowchart is compared with the actual, it shows points of deviations. A new process or plan can be tested for its logical consistency by following all paths of the flow chart.

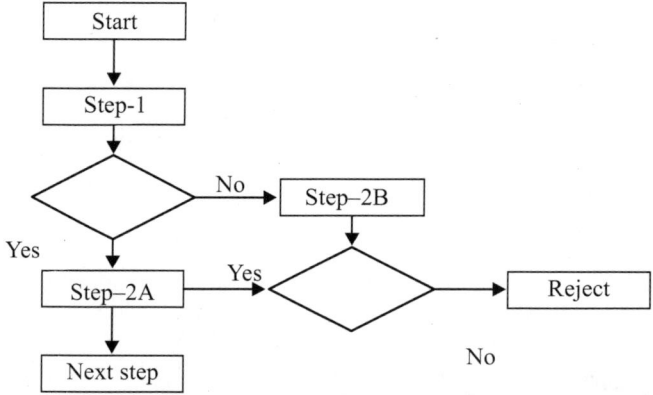

Figure 6.1 Typical flow chart.

When we get a problem, it is essential to analyse the path in which the work took place and verify the reports and the actions mentioned. Identify the possible reason for this problem to remain unnoticed during the process. If we have a habit of documenting all the events that took place while carrying out an activity, we shall be able to identify the problem in its root itself, and would not allow it to grow. In majority of the cases, we fail to identify the problem in the beginning because of lack of documentation. It holds well not only for the organizations, but also for the activities at home and society.

In a number of cases it is found that the poor quality or a problem was due to a short cut taken in the process, knowingly or unknowingly, and a critical check point was overlooked. Few of the examples are weighing the yarn received and issuing it to knitting assuming the weight written on the bags is correct, not verifying the wheels put before starting a spinning machine, issuing the materials to production without inspection, approving a payment without verifying the papers thoroughly, despatching the materials to customer without final approval of quality assurance, creeling the cones on warping without verifying the dyeing lot number and shade numbers, and so on.

6.6   Critical activity chart

After understanding a flowchart, information is gathered on each step to understand its criticality in process. The critical activity chart is a tool for systematically gathering and analyzing information about job process or operations, concentrating mainly on inputs, outputs, and basic processes, rather than specific interlinked sequence of process steps. This is used to understand the work process and define its boundary in problem identification stage, to help brain storming of problems in work place, and to identify major causes for dissatisfaction of internal customer to identify the major areas of output and their internal customers and assess the extent of their satisfaction. To prepare a critical activity chart, the major work activities are identified; and for each activity, one chart is prepared. The next step is identifying the inputs, outputs, tasks, and the customers. Then problems or deviations in output, input, and processes are identified. Problems can be clarified by discussing the contents with all concerned. It is a tool for common understanding. A good critical activity chart helps in highlighting the workplace activities and if analyzed systematically, makes brain storming faster and more useful.

INPUTS	WORK ACTIVITY	OUTPUTS
SUPPLIERS **External** **Internal**	TASK DETAILS	CUSTOMERS **External** **Internal**
PROBLEMS WITH INPUT	PROBLEMS WITH PROCESS	PROBLEMS WITH OUTPUT

Figure 6.2 Critical activity chart.

By identifying the critical activity and controlling it at that place not only makes the process effective but reduces the cost of operations and the overhead expenses for process monitoring. There is no need to spend on controlling of non-critical issues. Also, it shall not allow the focus to be diverted.

6.7   Boundary analysis

Boundary analysis, which is often referred to as districting, helps define regions according to certain criteria. A boundary analysis illustrates the relationships that exist within the processes. It details all interactions among processes, customer, and suppliers. First the starting and ending points of the process are defined. Then details are prepared for each process.

Customer input → Perceptions, requirements, complaints, expectations

Process output to customer → Information, deliverables (products/services)

Supplier input → Information, deliverables

Process output to supplier → Perceptions, requirements, complaints, expectations

Resources → Equipment, people, budget, procedure, training

Boundary analysis is appropriate in the exploratory stage and the hypothesis testing stage of research. During initial data exploration, boundary analysis can identify spatial patterns and generate testable hypotheses. Designing experiments for hypothesis testing requires more careful planning and a more thorough understanding of the analytical techniques to be used.

In case of yarn, by understanding the CV% of count, strength, etc., it is possible to work out the expected maximum and minimum readings, and also the probability of a certain reading coming again and again.

Scale of sampling: An important consideration in any spatial investigation is the scale of the sampling framework. By scale we mean both the size of the geographic area under study, and the spatial intervals at which observations are made. Ideally, the scale of the sampling regime reflects the scale of the processes under investigation. Determination of the appropriate scale may require a pilot study or other preliminary work. A sampling regime that is too broad or too narrow for the relationships under study will likely result in failure to detect boundaries or associations which may actually exist. In the event of non-significant findings, a logical first question is, "was the scale appropriate for this study?"

Take an example of a weaving factory purchasing yarns. The management does not want yarns to be wasted in the pretext of testing as the yarn is very costly. So, the quality assurance person takes one cone from each lot and checks for count and CSP. He takes five readings. If a lot is of 1000 kg, he takes sample from one cone which is in reality one cop, whereas there shall be around 20,000 cops in one lot. It is assumed that the remaining 19,999 cops shall be similar. In order to ensure that yarn is of right quality for the type of loom and the fabric being woven, we need larger samples.

Choice of variables: Within boundary seer, boundaries may be delineated based on one or many variables measured at a set of study locations. For example, in ecology, ecotones (boundaries between adjacent ecosystems) may be delineated, based on changes across space in the abundance of one dominant plant species, or based on changes in many plant species. The corresponding data sets would consist of data representing the abundance of plants measured within some unit of area at each spatial location. The first example would have only one variable for the focal species, while the second would have a column for each species sampled.

Selection of variables to include in a data set should start with existing knowledge of the system. Once a set of candidate variables has been constructed, a combination of techniques may be used to decide which variables are included in the boundary analysis. The first method is to look for boundaries for single variables, evaluating each variable independently. Then select variables for a multivariate boundary delineation, based on some predetermined criteria. For example, you may include only those variables that have significant boundaries themselves (determined using sub boundary analysis), or you may include those variables that have high rates of change in the same vicinity. An alternative method is to use multivariate techniques such as principal components analysis (PCA) to determine which of several candidate variables contribute significantly to the overall variation in the system. You might then decide to include variables that account for a certain proportion (e.g., 90%) of this variation. In any case, let the research question or process model, rather than models of data alone, guide selection of variables.

PCA is a mathematical procedure that uses an orthogonal transformation to convert a set of observations of possibly correlated variables into a set of values of linearly uncorrelated variables called principal components. The number of principal components is less than or equal to the number of original variables. This transformation is defined in such a way that the first principal component has the largest possible variance (that is, accounts for as much of the variability in the data as possible), and each succeeding component in turn has the highest variance possible under the constraint that it be orthogonal to (i.e., uncorrelated with) the preceding components. Principal components are guaranteed to be independent, only if the data set is jointly normally distributed. PCA is sensitive to the relative scaling of the original variables.

Making sense of boundary analysis: Boundary overlaps the statistics and address the question that "Are boundaries for two data sets significantly close to each other?" Implicit in this question is the assumption that the boundaries exist for the two suites of variables. Thus, boundaries must first be evaluated before assessing overlap.

For difference boundaries we suggest you evaluate this assumption by first calculating sub-boundary statistics for each data set. Sub-boundary statistics will assess boundary contiguity. If contiguous boundaries exist, then the interpretation of boundary overlap is clear: discrete boundaries overlap. If clear boundaries do not exist within each data set, yet overlap is significant, then the two suites of variables have a more complex relationship. In this case, areas of high rate of change for each data set coincide. Further investigation may be needed to uncover the nature of the relationship.

Boundary analysis might help in finding the reasons for variations in count CV% of yarn and shade variations between lots of yarns causing barre effect in knitted fabrics.

6.8   Check sheets

Check sheet is a well thought out format for collecting and compiling data for events as they happen that makes it easier for subsequent analysis. It may be a form, format, or table. The check sheets are useful in following areas:

(1) To understand the past and present status of the problem situation.

(2) To stratify the data as they are collected.

(3) To understand the change through the passage of time trend.

(4) To analyze the data as they are collected.

(5) To determine the details of defects.

(6) To determine where the defects occur.

(7) To inspect machines or equipments.

(8) To verify the operating procedure.

The check sheets are of three groups, viz. check sheets for recording data and making surveys, inspection and validation check sheet, and check drawings. The check drawings are helpful in finding the exact location of defects in order to identify problem area. Concentration diagram is a type of check drawing.

A checklist is a type of informational job-aid used to reduce failure by compensating for potential limits of human memory and attention. It helps to ensure consistency and completeness in carrying out a task. The preparation of checklist involves the following steps. It starts from clarifying the objectives by clearly stating the event or issue being observed and what data pertaining to it is to be collected. Depending on the type of the problem, one needs to decide as to when and where the data is to be collected and also the type of check sheet. Depending on the role of each item in the problem, the items to be checked are decided. First, a trial check sheet is prepared to ensure its suitability for collecting the data. The check sheet should include the title, object/items to be checked, checking method, date and time of check, the checker/observer, the location, and the summary of conclusions. While recording the observations, simple note using symbols can be made so that maximum information can be gathered in one stroke. The information collected is to be tallied for their completeness. Completely filled check sheet offers clearly visible data for the event, and it is self explanatory.

6.9   Concentration diagram

Concentration diagram is a special check sheet to record data about frequency, type, and location of events (defects or errors) on the picture or schematic drawings that are easily understood and visualized. It is used when visual picture or layout of location of event is more clearly understood than possible description; the possible locations are many and proper classification and expression by words are difficult, and when standard check sheet becomes difficult to understand for data collection in remote locations like the exact point of defect. The concentration diagrams when completely filled show the frequency and location of the event. The form is self explanatory, as it indicates the location in the diagram. This can be further analyzed by using other Q.C. tools.

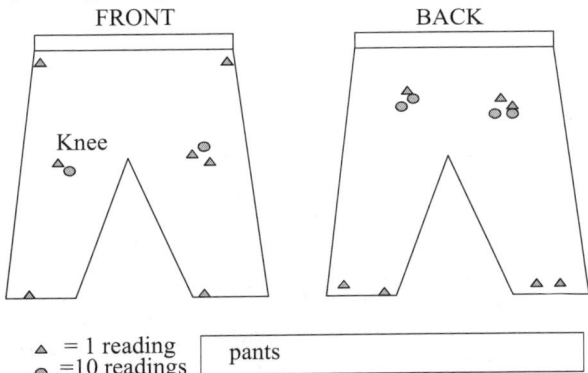

Figure 6.3 Concentration diagram – Wearing out of pants.

6.10   Stratification

Stratification means the separation of data into categories. This is a statistical technique of breaking down values and numbers into meaningful categories or classification to focus corrective action or to identify true causes. This is used to identify the category which contributes to the problems being tackled.

Graphs are among the simplest and best techniques to analyse and display data for easy communication. Stratified data is normally displayed in bar charts that show comparative characteristics by the length of the bar.

The stratification can be used:

• before collecting data.

• when data come from several sources or conditions, such as shifts, days of the week, suppliers, or population groups.

• when data analysis may require separating different sources or conditions.

Selection of stratification variables is essential for planning the variables and collecting data, rather than going on adding the variables as and when some information is obtained. For example, while collecting data for the reason of fabric defects, one can stratify them as warp yarn related, weft yarn related, winding related, warping related, sizing related, loom related, weaver related, and so on.

The cost and time for collecting additional identifying variables in the initial times is lesser as compared to the cost of new collection effort. The next step is to establish a category for each variable, which is a value or a range of values of a stratification variable. Then data is collected pertaining to those variables. The data obtained is sorted and grouped into stratified variables. If the first attempt of stratification does not reveal any significant pattern, data is collected again to find out the effect of other variables.

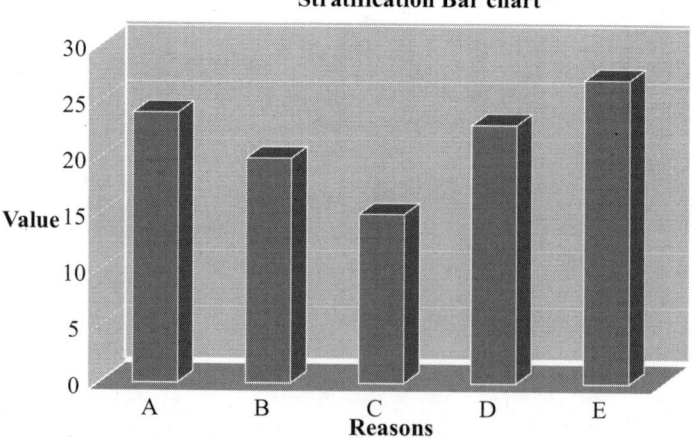

Figure 6.4 Stratification

Stratified run-chart: A run-chart instead of a bar chart is adopted for analyzing the system related problems. Let us take an example of a process and the time taken for doing different processes as shown below:

Time required for the process - minutes						
Process	A	B	C	D	E	F
1-March-07	20	43	25	32	25	12
2-March-07	21	40	25	35	26	14
3-March-07	20	35	25	31	24	16
4-March-07	21	43	26	30	25	17
5-March-07	23	30	25	28	26	13
6-March-07	21	26	26	42	27	15
7-March-07	20	40	25	35	25	16

For doing a job, a number of steps are involved and each step has its own importance and criticality. When we analyse the time taken for each element of a job, the normal practice is to collect data element-wise and project in a bar chart where either the average time or total time is shown. However, this cannot identify whether there is a real scope for improvement, or standardization, or a technological change is required. Normally, the tallest bar is taken as a target for improvement project. When we draw a stratified run-chart as shown in Fig. 6.5, we can identify the processes which have wide variations and the processes that are consistent. The processes with variations can be improved by standardization; whereas if a process is consistent, improvement can be achieved by technological change.

Stratified run chart

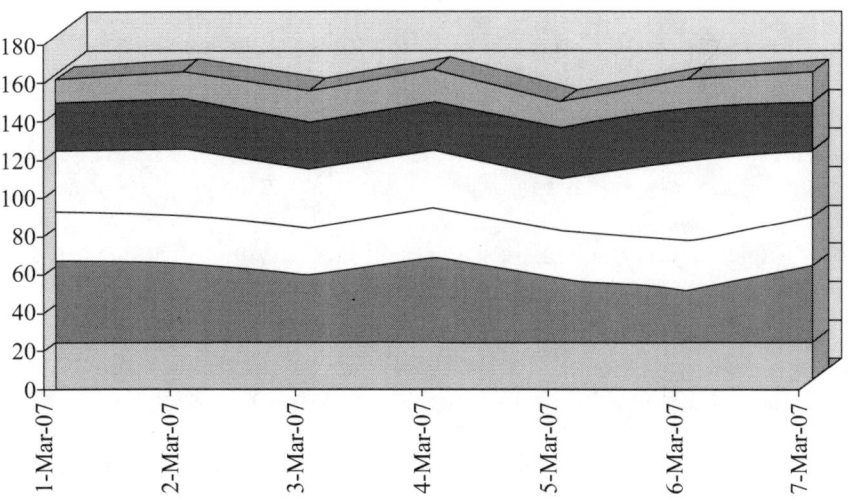

☐ Time required for the process A ☐ Time required for the process B
Figure 6.5 Stratified run-chart.

Stratified run-charts can be used for reducing the wastes, improving efficiency, implementing lean systems, and delays in maintenance activities, etc.

6.11   Run-charts and control charts

Run-chart displays the trend of changes of a character over a period of time. The X-axis always refers the time and Y-axis indicates the character under observation.

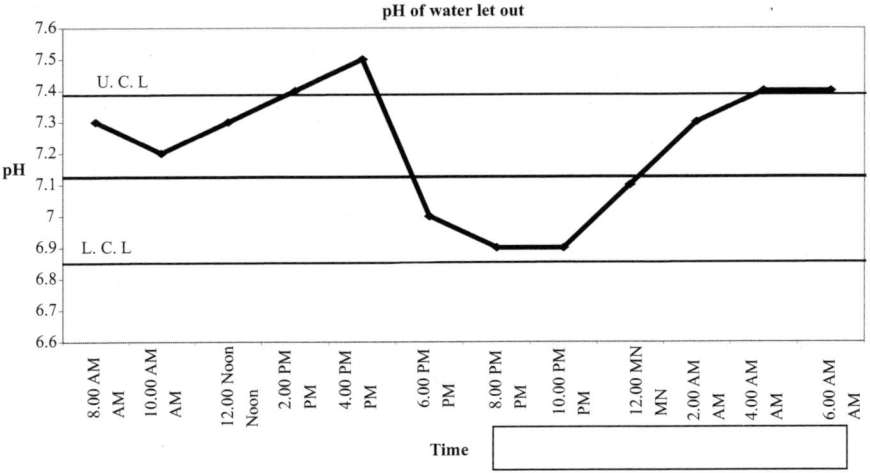

Figure 6.6 Control chart.

This is a specialized graph which uses connected lines instead of bars to illustrate data. A run-chart with statistically derived control limits is a control chart. This is used to highlight the variations of a characteristic over a period of time and to seek explanation for changes, to study the growing or declining trend of the average, to highlight significant improvements in performance after implementing counter measures, and to verify the effectiveness of the measures taken for improvement.

Control charts, also known as Shewhart charts (after Walter A. Shewhart) or process-behavior charts, in statistical process control are tools used to determine if a manufacturing or business process is in a state of statistical control. To make a control chart, time intervals are marked in the horizontal axis. The numeric scale must move in regular intervals. The vertical scale is marked while considering the expected range of variation. Control limits are shown on the chart, which is statistically worked out by considering the normal variation of the population. A typical control chart is shown in the Fig. 6.6 for pH of water let out from a process house. The control-charts are interpreted by identifying points in time when the characteristic changes significantly. If it is compared with other possible changes at the same time, it offers clues for causes. A common trend may be increasing or decreasing. The action is to be taken if the reading goes out of limits or continuously remains near any of the control limits.

6.12   Cause and effect diagram

Cause and effect diagram is a representation of the systematic relationship between the event under investigation and all possible causes influencing. It is also a documentation of group thinking process to investigate the root cause of the event. It looks like a skeleton of a fish known as fish bone diagram and also as Ishikawa diagram in the name of its founder. This is used to investigate the cause and effect, and helps stratification for collection of data to confirm relationship and evolve counter measures. The steps involved are defining clearly the problem, or effect, or event for which the cause is to be identified.

Common uses of the Ishikawa diagram are product design and quality defect prevention to identify potential factors causing an overall effect. Each cause or reason for imperfection is a source of variation. Causes are usually grouped into major categories to identify these sources of variation. The categories typically include the following:

- People: Anyone involved with the process.
- Methods: How the process is performed and the specific requirements for doing it, such as policies, procedures, rules, regulations, and laws.
- Machines: Any equipment, computers, tools, etc., required to accomplish the job.
- Materials: Raw materials, parts, pens, paper, etc., used to produce the final product.
- Measurements: Data generated from the process that are used to evaluate its quality.
- Environment: The conditions, such as location, time, temperature, and culture in which the process operates.

A horizontal line with an arrow at the right hand end and a box in front of it is drawn. Problem statement is written in the effect box. The next step is identifying the causes in major categories. Brain storming is normally used to identify the possible major causes. After identifying a primary cause, the team shall go in deep and identify as many secondary or tertiary causes as possible in each of the primary causes. Each of the major causes are placed in a box horizontal to the first line and connected to that line at an inclination of approximately 70°. After identifying the major causes, the root causes are investigated by adopting root cause analysis techniques. The logical validity is checked for all causes identified, considering the present scenario. It is important to understand the potential pit falls while using cause and effect diagram. It should not be treated as a substitute for data. It should be drawn only after preliminary data has been collected to narrow down the focus of a problem. One should not limit himself just to those theories that are in the diagram.

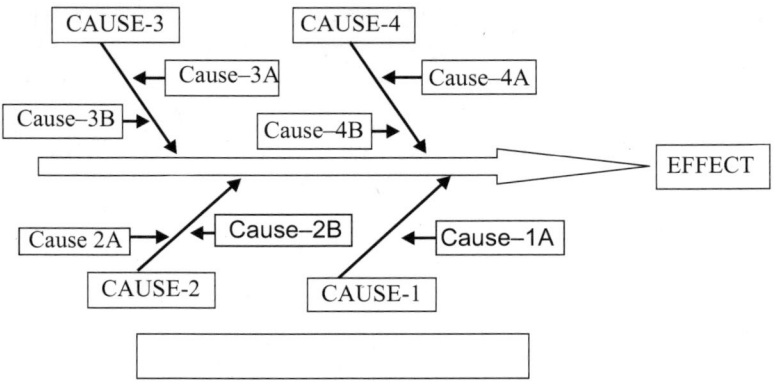

Figure 6.7 Cause and effect diagram.

6.13   Pareto analysis

Pareto analysis is a statistical technique in decision making that is used for selection of a limited number of tasks that produce significant overall effect. It uses the Pareto principle – the idea that by doing 20% of work, 80% of the advantage of doing the entire job can be generated. Or in terms of quality improvement, a large majority of problems (80%) are produced by a few key causes (20%). A Pareto diagram is a special form of vertical bar graph that helps in identifying "vital few" from the "useful many". Its concept was given by Mr. Wilfred Pareto, an economist from Italy, and was developed as a Q. C. tool by Prof. J. M. Juran. The principle involved is that very few causes contribute for maximum effect; whereas a number of other factors contribute only for a small effect. It is used while setting priority, while selecting the problem, and for identifying the most important root causes contributing substantially to the problem.

The impact of each factor is worked as a percent of total impact, and the factors are arranged in a descending order. A bar chart is prepared. In addition to this, a line graph is prepared for the cumulative impacts worked starting from the highest contributing factor.

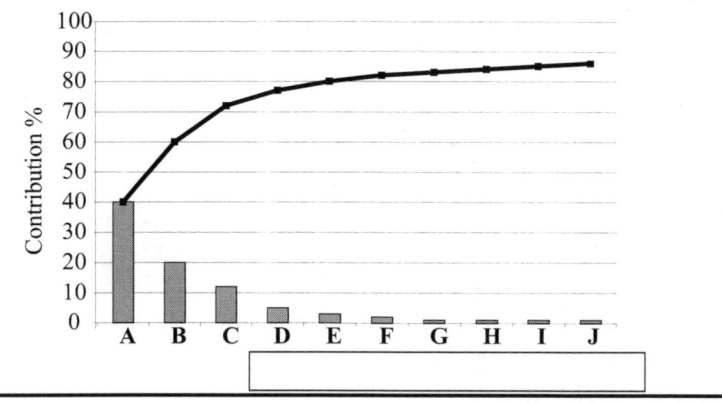

Figure 6.8 Pareto analysis.

6.14   Histogram

A histogram is a special type of bar chart to show the distribution or spread of the observed characteristics, which enables one to see patterns that are difficult to see in a simple table of numbers. It is a visual presentation of range, magnitude, central tendency, and the spread. Histogram was first developed by French statistician A. M. Guerry in 1833. The histogram helps to identify whether a spread is normal or not, the surprises in the natural distribution which can lead to causes or counter measures, confirm the results of a process improvement, and to obtain clues for stratification. The common histogram patterns are normal bell shaped distribution, double peaked distribution, plateau distribution, comb distribution, skewed distribution, truncated distribution, and isolated peaked distribution or island distribution.

6.15   Scatter diagrams

Scatter diagram is a simple graphic presentation of the relationship between two variables, which relates to cause and its effect. It is one of the oldest applications of graph, and was developed by Dr. Burton in London in 1794. In 1832, Mr. J. F. W. Horshel fitted a curve to the scatter diagram. It is another quality tool that can be used to show the relationship between "paired data", and can provide more useful information about a production process. A scatter diagram, also called as scatterplot, or scatter graph is a chart that uses Cartesian coordinates to display values for two variables. The data is displayed as a collection of points, each having one coordinate on the horizontal axis and one on the vertical axis. A scatterplot does not specify dependent or independent variables. Either type of variable can be plotted on either axis. Scatterplots represent the association (not causation) between two variables.

The points marked on a scatter diagram form a pattern which indicates the degree and nature of relationships, which is statistically known as correlation.

Scatter diagrams are used to verify if there is any relationship between cause and effects with facts, and to estimate the strength and nature of the relationship between two sets of data. The concept is that there is always a relation between a cause and effect, but it would be difficult to state them in precise mathematical terms. It is easier to see the relationship in a scatter diagram than in a simple table of numbers. The effective problem solving is possible only when we discover and test the true relationship between a cause and its effect. Normally, the suspected cause is taken in X axis and the effect in Y axis. Points are plotted for each reading of cause and effect. The pattern generated by the cluster of points gives clue to the possible relationship.

The scatter diagrams indicate the relation as strong positive, strong negative, weak positive, weak negative, and no relation. If the pattern of dots slopes from lower left to upper right, it suggests a positive correlation between the variables being studied. If the pattern of dots slopes from upper left to lower right, it suggests a negative correlation. A line of best fit can be drawn in order to study the correlation between the variables. An equation for the line of best fit can be computed, using the method of linear regression.

There may be few complicated relations where the value of Y increases as X increases to a certain extent, and then it might change its direction. The relation need not be linear all the time. Depending on the degree of relationship, further statistical analysis or verification can be carried out. Also, quantification of degree of relationship can be carried out by working out coefficient of correlation.

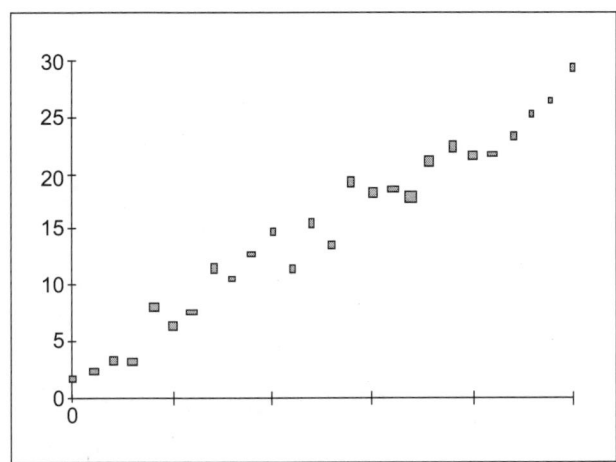

Figure 6.9 Strong positive correlation.

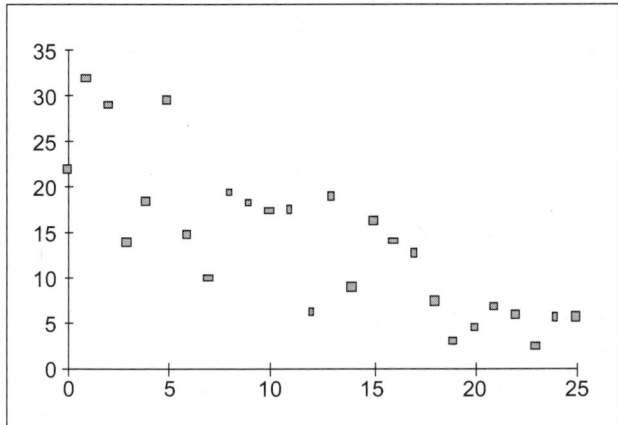

Figure 6.10 Moderate negative correlation.

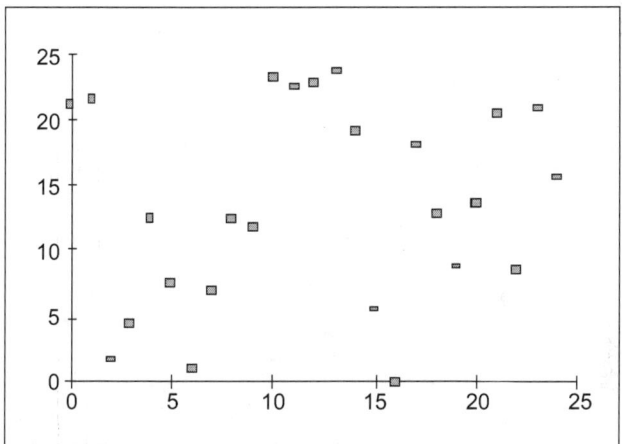

Figure 6.11 No correlation.

In the Figs. 6.9 to 6.11, the dots which are actually data points have various relationships. The strong correlation indicates that there is a close relationship between the data that is paired together. In the middle diagram, you see a slightly different pattern indicating that there is, in few cases, a relationship and in other cases there is no relationship. The last diagram on the right indicates that there is no correlation, or no relationship at all between the paired data.

In the Fig. 6.9, you would be able to determine that you have a strong relationship and thus one measurement has a strong relationship to the other; therefore, you would be able to prove that one item affects the other closely.

In the Fig. 6.11, you would be able to determine that there is absolutely no relationship between the two items, and you need to review the "cause-and-effect" diagram, or "brain-storming" session to try and find another item that your primary item measured, might have a relationship to.

The Fig. 6.10 is the one that is going to cause you some grief. This particular diagram is more difficult to interpret, and actually requires a more detailed investigation into which data points correlate, and in which data points have absolutely no comparison. Then, you need to try and determine why certain ones reveal a relationship and others do not.

The most common mistake with regards to scatter diagram is the failure to use them, as people assume that a relation exists and need for showing it diagrammatically is not felt. Sometimes, we get correlation without physically understanding the reason or relationship.

The scatter diagram is a useful tool, but it cannot substitute the team's knowledge and fundamental understanding of the process and the problem under study.

6.16   Force field analysis

In the improvement process, if the improvement is to be successful, few changes have to take place. There are some hindrances for the change and some elements support the change. Force field analysis is a technique developed by Kurt Lewin to identify elements which resist the change (hinder) and which are pushing for change (aids). This helps in developing the implementing strategies for a change by carefully working with the factors which favor or hinder the process. This is used for chalking out possible implementation strategy for an improvement, to forecast and assess the problems likely to occur from hindering factors while implementing a change, and to develop counter measures to minimize the impact of hindering factors during successful implementation.

If the helping factors are more powerful, the change takes place and implementation of counter measure is successful. Normally, each hindering factor has a counter factor that can help. Force field analysis should be analysed to find these couples. If this is not obvious, it is possible to generate additional 'drivers' to facilitate implementation. Then if helping factors are nurtured, implementation has better chances of success.

Force field analysis is best carried out in a small group of about six to eight people, using flipchart, paper, or overhead transparencies so that everyone can see what is going on. The first step is to agree the area of change to be discussed. This might be written as a desired policy goal or objective. All the forces in support of the change are then listed in a column to the left (driving the change forward), whereas all forces working against the change are listed in a column to the right (holding it back). The driving and restraining forces should be sorted around common themes and then be scored according to their 'magnitude', ranging from one (weak) to five (strong).

Throughout the process, rich discussion, debate, and dialogue should emerge. This is an important part of the exercise and key issues should be allowed time. Findings and ideas may well come up to do with concerns, problems, symptoms, and solutions. It is useful to record these and review where there is consensus on an action or a way forward. In policy influencing, the aim is to find ways to reduce the restraining forces and to capitalize on the driving forces.

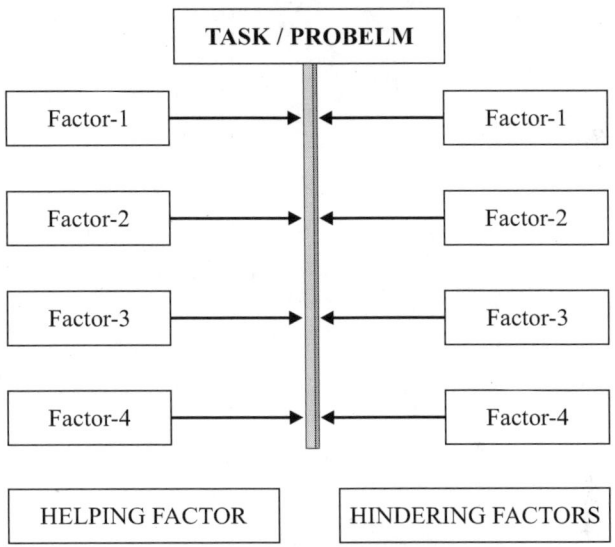

Figure 6.12 Force-field analysis.

Let us take an example of considering better cotton for spinning a given yarn than the existing one. The spinners prefer better cotton as it increases productivity, reduces saleable wastes, improves performance in winding, warping, weaving, or knitting due to lesser unevenness and higher strength, improves feel of fabric, lesser dust liberation while spinning and weaving, and so on. The management, although agrees for all, argues that the customers are not ready to pay any higher price because the yarn is good but insist on the performance in the same level if we give good yarn to them. The customer always refers the lowest price at which yarn is available in the market, and insists us to quote lower than the lowest quote to enable us to sell the yarn in the market. The cotton prices are not stable, and we cannot always maintain the same price.

The following diagram explains the favouring and hindering factors for the decision to be taken on the cotton to be procured.

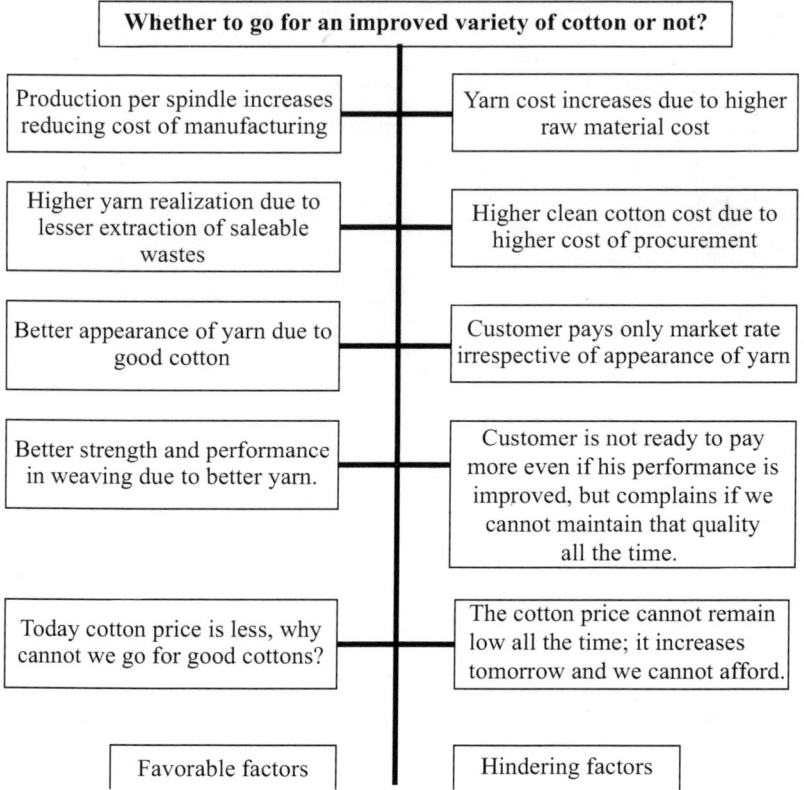

Whether to go for an improved variety of cotton or not?	
Production per spindle increases reducing cost of manufacturing	Yarn cost increases due to higher raw material cost
Higher yarn realization due to lesser extraction of saleable wastes	Higher clean cotton cost due to higher cost of procurement
Better appearance of yarn due to good cotton	Customer pays only market rate irrespective of appearance of yarn
Better strength and performance in weaving due to better yarn.	Customer is not ready to pay more even if his performance is improved, but complains if we cannot maintain that quality all the time.
Today cotton price is less, why cannot we go for good cottons?	The cotton price cannot remain low all the time; it increases tomorrow and we cannot afford.
Favorable factors	Hindering factors

6.17 Spectrograms

Spectrogram is a QC tool being used in textile mills to locate the source of fault in a yarn, filament, rove, sliver, or any such continuous strand that are produced by using rotating rollers. It highlights the defects which occur in a regular frequency. By carefully studying the gearing diagram and working out backwards, it is possible to pinpoint the source of defect. This tool can be used in manufacturing industry producing drawn wires, where a pair of drawing rollers draws the wires by applying certain draft.

The curved line is an ideal curve, whereas the actual values, i.e. the frequency of a particular type/size repeating are shown as blue bars. If the height is more than the ideal curve, then it needs to be corrected. However, the readings that are more than twice the height of ideal curve are considered as significant. In the Fig. 6.13 the highest peak is at 8 cm, indicating a defect occurring at every 8 cm of the product. If the draw rollers have a diameter of 25 mm, then for every one revolution we get $25 \times 3.14 = 7.9$ cm. It means there is a problem in the draw roller. Similarly, we can work out the source of defect by understanding the gearing and working out the length of material produced for one revolution of each roller or gear.

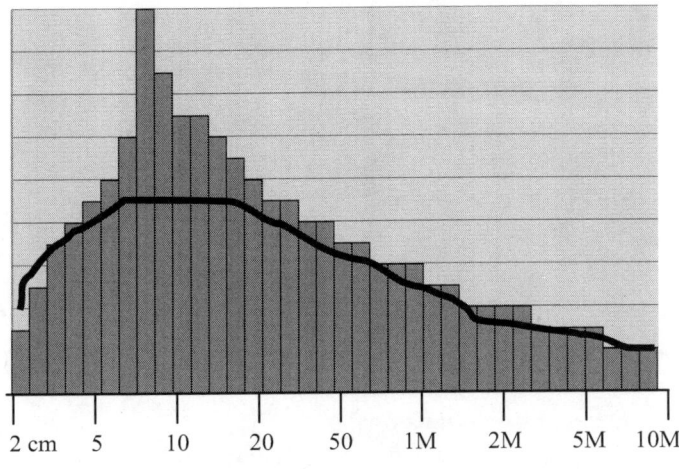

Figure 6.13 Spectrogram.

Now softwares are available for analyzing spectrograms. You need to enter the wheels that are running in your machine, and the diameters of the rollers, and keep it in memory for the type of machines you are using. Then enter the reading at which you are getting peaks, and then you get the wheel or roll causing that defect.

6.18   Listing pros and cons

If an established set of criteria already exists, the evaluation of the options is simple, by using comparison tables, with all criteria of equal weight. However, it is more likely that a situation is not that simplistic with little or no clear criteria, like deciding what you should do next from a set of unrelated possibilities. Mycoated recommends the use of the pros and cons approach with only 2–3 options; lists the pros and cons for each and compares the results directly. However, working with larger number of options requires the following more systematic approaches:

(1)   Generate a comprehensive collection of pros and cons by working through the options one by one, and generate a realistic set of pros and cons for each using creativity approaches. Write each pro or con on a separate card or post-it, clearly marked '+' (for a pro) and '−' (for a con).

(2)   Collate the collection into an ordered checklist of criteria with pros and cons stacked separately, any duplicates removed and a single master checklist of all pros and all cons prepared. Focus on the central issue you are working on and order the lists Vital (make or break), Important (but not absolutely vital), Marginal (i.e., would be nice if...). These categories can be sub divided further if necessary.

(3) Pick out 'vital' options by making a 'short-list' of potentially viable options. If unsure about an item, do not exclude it, yet.

(4) From the 'vital' short-list, pick out 'important' options, counting the number of 'important' pro criteria that are present, and con criteria that are absent. Eliminate all options that score poorly at this stage, to leave a list of feasible, good quality options.

(5) Repeat with the 'marginal' criteria, condensing the short-list yet further to only options that are feasible, of good quality, and which have useful additional properties.

(6) This technique is used mainly for screening out clearly weaker options, using vital/important/marginal distinction. It does not make finer distinctions within a final short-list.

Once we are clear of using different tools available depending on the situation, we can start our journey of diagnosis and remedial actions to achieve our goals by solving the problems.

6.19   RPR problem diagnosis

RPR (rapid problem resolution) is a problem diagnosis method specifically designed to determine the root cause of IT problems. By carefully studying the concepts, the same can be used for number of repeated technical problems in textile industry. As number machines are now PLC driven, this technique shall be handier. RPR deals with failures, incorrect output, and performance issues, and its particular strengths are in the diagnosis of ongoing and recurring grey problems. The method comprises of the following:

- Core process
- Supporting techniques

The core process defines a step-by-step approach to problem diagnosis and has three phases as follows:

(1) Discover
   - Gather and review existing information
   - Reach an agreed understanding

(2) Investigate
   - Create and execute a diagnostic data capture plan
   - Analyse the results and iterate, if necessary
   - Identify root cause

(3) Fix

- Translate diagnostic data
- Determine and implement fix
- Confirm root cause addressed

The supporting techniques detail how the objectives of the core-process steps are achieved, and cite examples by using tools and techniques that are available in every business.

RPR has few limitations and considerations, including the following:

- RPR deals with a single symptom at a time.
- RPR identifies the technical root cause of a problem, and it can't be used to identify the non-technical root cause with people, process, etc.
- RPR is not a forensic technique and so historical data alone is rarely sufficient.
- The investigate phase requires the user to experience the problem one more time.

7.1 Diagnosis

Diagnosis is to ascertain the cause or nature of a disorder, malfunction, problem, etc., from the symptoms. To cure a disease, proper diagnosis is very important by understanding the root causes and focalizing thinking around solutions that can truly bring us closer to where we wanted to be. Arnaud opines that it is a good idea to first make sure that we understand precisely the problem before looking for ways to correct it.

Diagnosis is to carry out the detailed analytical work needed to identify the root causes of chronic problems. The process starts from the symptom and is often difficult as compared to the remedial action. In the majority of cases, a chronic problem shows its symptom in one department or area; while the cause lies in a different section or department. In such cases, cross functional teams shall come to help.

Take the case of textile industry in India. The employee turnover is very high, and the HRD persons have to be always on their toes to bring new employees. The actual root cause may be lying in the production department where the working conditions are not good, or the boss is demanding more from the people than their capacities, or ill-treating subordinates, whereas the problem is referred as HR problem and head of HR is held responsible. Similarly the short sighted decisions taken by top management might have created problems at a number of places; the person suffering or reporting the problem is held as responsible for the problem.

Direction and diagnosis are the two processes in a diagnostic journey. The direction provides the project definition and various theories that are to be tested. The theories are outcome of the logical thinking, previous experiences, and the knowledge of the members involved in diagnosis. When theories are tested, a numbers of them turn out to be invalid; whereas some shall emerge as valid and become the basis for subsequent remedial action.

7.1.1 Steps involved in diagnosis

The steps involved in diagnosis are: analyzing the symptoms, translating theories into data requirement, testing the theories, analyzing and summarizing the results. The persons involved in diagnosis must have time, diagnostic skills, objectivity, and a factual approach. The time is needed to carryout numerous tasks, viz. precisely defining each symptom in order of frequency, applying Pareto principle, designing a plan of data collection and analysis, carrying out data collection, conducting data analysis, summarizing, and presenting the results. The diagnostic skills are concerned with the scientific testing of theories, designing the plan for data collection, collecting the data without bias, and bringing the meaning out of the resulting data.

Chronic problems continuing for years give rise to long-standing biases, and it is difficult to make the people understand that a problem exists. The factual approach helps in eliminating biases. Hence, data is the lighthouse for diagnostic journey.

In case of diagnostic skills, there is usually a remarkable difference in the skill level demanded for breakthrough as compared to that demanded for troubleshooting. In troubleshooting, the objective is to restore status quo; whereas in breakthrough, it is to reach a level of performance never attained before. It is a voyage of discovery, and the key question is to identify the variables that stand between us and getting rid of this chronic COPQ (Cost of Poor Quality). In order to get there, we need a map or a procedure which lays out the route. The diagnostic journey consists of the following phases:

- Analyzing symptoms
- Formulating theories
- Testing theories
- Identifying the root causes

One of the important concepts for effecting quality improvement is the concept of controllability. Studies are used to learn if errors are worker-controllable or management-controllable. The analyses of errors in most controllability studies indicate that over 80% are management-controllable and another 20% are worker- controllable. An error is worker-controllable if workers have the means of knowing what they are supposed to do, means of doing what they are actually doing and have means available to them for regulating the performance.

After identifying or understanding the symptoms, the next step shall be formulating and arranging theories. The progress in diagnosis is made theory by theory, by affirming or denying the validity of these theories about the causes. The theories are to be identified by brain storming, and it is necessary to record all theories without challenging them. The best source of theories is the men on job. They might have even done an experiment to get rid of the problem. As the list of theories grows, it is useful to arrange it in a way which shows the interrelation of the various theories. Ishikawa diagram can be used to identify the relation and interaction of theories.

After listing the theories, next step is to test the theories. Which theory should be tested is decided by the experience of managers. Few of the theories can be tested readily by using data already in house; whereas few might involve pains-taking experiments, which if designed improperly are likely to end in frustration.

There are four broad methods of approach for testing theories. They are as follows:

- Using existing data

- Using current operations data

- Cutting new windows

- Designing and conducting experiments

It is suggested that the mills collect data and preserve them properly so that they can be used for analyzing the problems. One of the biggest problems found in textile and garment industry is the lack of clarity as to which data are to be collected and preserved. Further question is how long to keep the data and how to keep.

If data is not available for analyzing a problem, then we need to collect data from the present operations and try whether we can get the root of the problem.

If a problem is chronic, it indicates that the data collected and analyzed are not addressing the root of the problem, but trying to attack some symptoms, and corrective actions are taken. For example if there is a shortage of workers, people just engage previous shift people on overtime and run the show. If working is not good in spinning, additional help of doffers is given or the speed is reduced. People normally do not make an effort to find out why the situation changed and what the root cause of it was.

Only by referring to the data recorded or observing what is happening, we may not get to the root. We may have to think in a different way, which was never tried, and it is referred as cutting new windows.

Designing of experiments should be done with logic and not by trial and error. Before making an experiment, one should have jotted down the possible results because of the factors you are considering. In number of cases, we do not get the anticipated results as we have not addressed all the possible variables. We may not be aware of some of the variables that are influencing our activity.

7.1.2 MECE thinking

Arnaud suggests building a "Logic Tree" for complex problems and a "why" tree to organize and test the hypotheses. This requires one to have a logical thinking process, and, in particular, need to think with mutually exclusive and collectively exhaustive (MECE) elements. Building logic trees also requires one to be innovative in thinking, going further than the obvious answers.

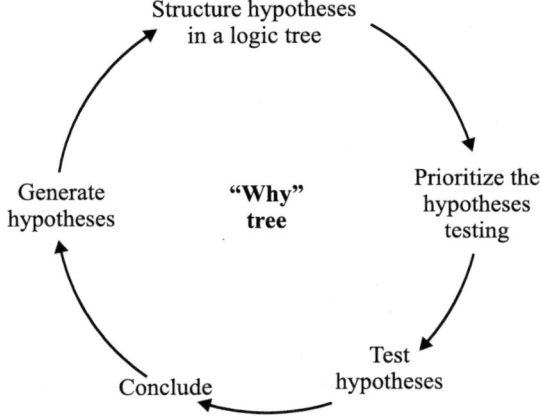

Figure 7.1 Logic tree.

A useful tool for building "why" trees and applying MECE thinking is to think in terms of processes. Another useful tool is to use existing frameworks whenever possible.

As with defining problems, you can't expect to be excellent at diagnosing problems without practicing them extensively. When you do, it is a good idea to enlist others to challenge your thinking.

Arnaud suggests avoiding explaining the process in "How" trees. There are two ways to answer to a "how" question. The first is to describe one particular solution process to reply to the question (e.g., to the question "how to clean a cot in ring frame?", reply: first, press the lever in top arm; second, lift the top arm up; third, take out the cot with arbour; fourth remove the lapped cotton by hand, never use a knife or any sharp object, etc.). The second is to describe the various alternatives in which one can answer the question (e.g., by keeping additional set of cots, by keeping the spindle working and cleaning the cot in leisure time, using solvent such as petrol to remove the sticky honey dew, coating with varnish to avoid sticking, acid treating, etc.).

"How" trees don't describe processes, instead they spell out the various ways to answer to your key question, organizing these solutions in a MECE way. "Why" trees can benefit from thinking in terms of processes and their steps. Since the problem occurs because at least one of its part doesn't function properly, all one has to do is to break the process in MECE steps before testing each step. For instance, suppose that you want to understand why the parts that you ordered don't come on time. Map out the process as a collection of successive MECE steps and these will be the structure of your issue tree. You can then refine each of these steps by mapping out their respective components into further detail.

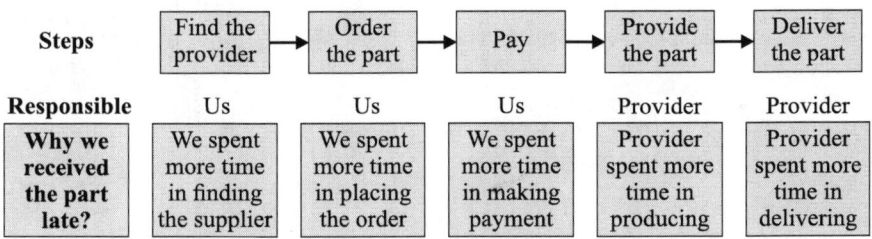

Steps	Find the provider	Order the part	Pay	Provide the part	Deliver the part
Responsible	Us	Us	Us	Provider	Provider
Why we received the part late?	We spent more time in finding the supplier	We spent more time in placing the order	We spent more time in making payment	Provider spent more time in producing	Provider spent more time in delivering

Figure 7.2 Using MECE thinking in "Why" tree.

MECE – Mutually Exclusive and Collectively Exhaustive thinking is explained as follows by Arnaud.

Two sets of elements are mutually exclusive when they don't intersect; you cannot have an element belonging to both sets at the same time. When you are mutually exclusive in your approach, you consider each potential solution only once; hereby ensuring that you do not duplicate efforts. Mutually exclusive thinking forces you to consider the details, seeing the individual tree as opposed to the forest. It helps you ensure that each element is different than the others.

Collective exhaustive means 'no gap'. Groups of solutions are collectively exhaustive when, in between them, they include all the possible answers to your problem. When your analysis is collectively exhaustive, it includes all possible solutions at least once. Collectively exhaustive thinking helps you ensure that you do not forget possible solutions; that is, you must be innovative, viewing the forest as opposed to its individual trees.

Being MECE will drastically improve your thinking. MECE thinking is perhaps the most important concept in analytical problem solving. The concept is simple to understand but it can be challenging to apply in some fields. That's because we're usually better at either considering minute details or the big picture but not both, let alone at the same time. Becoming a strong MECE thinker takes training and you should make it a habit to think in MECE ways. Making it a habit means that, each time you are confronting a new problem, you need to actively look for a MECE way to break it down in its root causes and/or solve it. All your issues trees have to be MECE

7.2 Remedies

Remedy is something that corrects an evil, fault, or error. After the various theories are tested, the resultant data will point to the possible cause. Once the cause is known, it is relatively easy to arrive at a consensus on effective remedial action. The journey ends when the remedy has been demonstrated as capable of producing improved results under operating conditions.

To ensure that the chosen remedy is optimal from the company's prospective, and it is successfully implemented, the following guidelines are to be used:

- Choose from alternative
- Anticipate resistance to change
- Implement remedial action
- Establish control at the improved level
- Suggest where the remedy can be replicated
- Feedback repetitive problems to product and process planners

There may be different alternative solutions for a problem, and one needs to choose from the alternatives by considering the cost involved, ease for implementation, acceptability to implementers, the probable side effects, etc. The necessity of making conciliation occurs every time an idea or innovation needs assessment. Suppose a new system has been thought up to meet a problem or situation which needs improvement, the question before getting into it shall be – "will that pay off? Who all will be benefited? Who all will suffer? Who is likely to support? Who is likely to resist?" before making a detailed planning of such a system. This question needs answering before much time, effort, and money are spent.

Let us take an example of a textile mill problem. Shortage of skilled workers is a problem. What are the alternate solutions suggested by the team?

(1) Pay more wages then people will come to you.

(2) Introduce good increment scheme so that people with long working years are paid substantially high, compared to people with less working years in the company.

(3) Provide living quarters to the workers near the mill.

(4) Provide schooling facilities to the children of employees.

(5) Introduce schemes of education fees reimbursement for workers and their children working for long.

(6) Provide employment to the kit and kins of employees.

(7) Provide housing loans for employees with long service.

(8) Provide good working atmosphere.

(9) Treat employees with dignity; do not ill-treat or use abusing language.

(10) Provide facility for employees to grow within the organization.

(11) Identify the real hard working people and encourage them; do not go by the words of your so called "close associates" [In Hindi, they are termed as "Chamcha" (चमचा). In number of companies it is seen that the main culprits are the people who are closely associated with top management, giving them advices and tilting the information as per their advantage.]

(12) Plan the activities well in advance and do not always insist as "Urgent".

(13) Take workers into confidence while making changes.

(14) Introduce concepts and systems of quality circles.

(15) Have transparent administration.

(16) Listen to the grievance of employees and take timely appropriate action.

(17) Bring large orders and allow for stable working.

(18) Do not go on making changes in production pattern.

(19) Educate the heads of the departments to treat their subordinates in a decent way.

(20) Allow the workers to form their union or association. Where there is no union, workers are not confident of their continued service and would like to join a company where there is security.

(21) Do not violate the government rules.

(22) Have ethical work practices.

Once the remedy is accepted, it should be implemented by the line department, but it is observed that the team encounters delaying tactics or outright rejection of the remedy. The source of resistance varies, which may be a manager, a supervisor, the workforce or the union. Always there are reasons to resist and surprisingly in a number of cases, such reasons for rejection are advanced by the very people whom the change is intended to benefit.

We normally deal with two changes, viz. a technological change and the consequences of the technological change. The social consequence is the trouble maker, which is a sort of uninvited guest and rides on the back of any technological change. A systematic approach of taking the people into confidence, starting small, transparency and no surprises, choosing the right time and moving with the culture of the society are helpful in a remedial journey with least resistance.

Implementing remedial actions includes providing a capable process to the operating forces that can hold the gains, establishing operation standards and procedures to serve as a basis for training, control and audit, providing training to use the procedures and to meet the standards and developing a system of control for detection and correction of out-of-control conditions. Establishing control at improved level is very essential to hold the gains. This might be done by process/operations audit and financial controls.

Replication of remedies for a similar quality problem elsewhere by suitable communication is essential to get the full advantage of the remedial journey. The design review should verify the remedies taken for various problems earlier, so the process can be designed properly with lesser problems.

A proper diagnosis and remedial action can help in solving the problems from their root, and lead to permanent solution and improvement. Hence, one can be a winner.

7.3   SODA – (Strategic Options Development and Analysis)

The SODA (Strategic Options Development and Analysis) was developed in the late 1980's. It is a methodology for helping one to understand the various viewpoints of a problem area. The general steps are explained below, whilst the detail are tailored to the specific problem.

(1) Planning meetings: The meetings are planned where the project is set up to have an initial view of the problem and the situation. At this point, it is important to decide who the participants will be and what the outputs will be in order to manage expectations.

In a number of cases, the mills fail in implementing their strategies because the decision is taken by someone and not communicated to people down the line on how to act. Also, the views of the people implementing and the problems they face are not discussed. The people implementing are not cleared of the doubts they had.

(2) Client interviews: Here the key people involved with the issue are interviewed, in a relaxed format, for an hour or so to obtain their individual views of the problem area and the situation. This is a very important step as the success of the implementation depends on the clarity the people have regarding implementation. Discussing in a relaxed manner is the key, as the real problems can be explained when the people are in a relaxed mood and not in tension.

(3) Development of causal maps: Causal mapping is used to get depicted the interviewee's perception of the situation. A causal map is a type of concept map in which the links between nodes represent causality or influence. Causal mapping is the process of creating a causal map (When done by an individual to clarify their own thinking, it is referred to as "cognitive mapping". It may be called "oval mapping" when done by a group, named after the small oval pieces of paper containing each idea).

The exercise of causal mapping helps in identifying the likely problems or hindrances much in advance, so that an action plan can be prepared for overcoming them. Much of the failures are because the people are not prepared to face the unexpected situation.

(4) Check-back interviews: Checking with the interviewees is done to ensure that the causal maps have correctly interpreted their views. If not, they are modified until they are a true representation.

(5) Merging the maps: The individual maps are combined to form a single map showing all the activities of the project at one place.

(6) Presentation: Both the individual and combined maps are presented to the participants, and the merged map is worked on until everyone finds it acceptable. This allows the whole group to understand all the viewpoints and to have ownership of the final map.

(7) Interpret the map in terms of goals, strategies, and tactics: The completed and agreed map can be used to determine the following:

- *High level goals* – These are usually where the causal arrow-heads that emerge from the map but don't go any further.

- *Medium level strategies* – These are generally the factors that feed in more or less directly to the goals.

- *Low level tactics and operational targets* – These are typically the activities that feed into the medium level strategies. They are often located at where causal arrows tend to come in from the wider environment.

(8)   Action selection, allocation, and implementation: Now that the goals, strategies, and targets have been determined; these need to be allocated to people for implementation.

       While allocating the activities, one should understand the influence that the person has in that area, so that he gets support from the people involved, and the implementation can be smooth.

7.4   Potential problem analysis

Potential Problem Analysis is a method designed by Kepner and Tregoe as part of their problem-solving technique. Its aim is to provide a challenging analysis of an idea being developed or action plan so that one can determine ways in which it may go wrong. It makes use of bulletproofing and negative brainstorming techniques. This method is used for identifying potential faults in complex systems. It has a "rational" rather than "creative" approach, but it still provides first-rate supply of creative triggers if approached in an imaginative spirit. A substantial amount of effort is required to carry out the analysis thoroughly and therefore, the method is usually set aside for the more ultimate action plan. This starts with defining the key requirements which are a "must". They may be actions, outputs, or events that must take place if the implementation is to be successful. Failure of any of these is likely to cause problems.

(1)   Record and investigate all possible problems for each of the key requirements that have now been identified, listing all 'potential problems', i.e. potential ways in which it could go wrong (a technique such as negative brainstorming could help) and look at each of them (a technique such as five Ws and H could help). If you have come up with more number of possible problems, it is advisable to make an initial estimate of the risk that each problem creates, so that you can give attention to the rest of the analysis on those that offer the greatest risk.

(2)   List possible causes for each potential problem, and the risk associated with it. The risk reflects both the likelihood of an event and the severity of the impact if it did, so that "high likelihood/high impact" causes present the highest risk.

(3)   Develop preventative actions where possible, rather than having to muddle through a problem after it has happened. Where possible, try to develop ways of preventing potential problem causes or minimizing their effects and estimate the residual risk that might still remain even if preventative actions were taken.

(4) Develop contingency plans where necessary, i.e. where problems would have serious effects, but you cannot prevent them, or there is a high residual risk even if you do.

(5) The table below is a simple way of displaying the analysis, which could contain a variety of quantitative estimates from crude 'high, medium, and low' subjective judgments, to carefully, researched measures depending on the demands of the situation.

Analysis for key activity: Complete project report for client					
Potential problem	**Possible causes**	**How likely?**	**Ways to limit risk**	**Residual risk**	**Contingency plans**
A: Report not delivered in time	Not prepared in time	High	Switch preparation to the 'A' team	Low	Allow generous margin in promised delivery time
	Mailing delays	Low	Hand delivery instead of internal mail	Minimal	Not needed – risk acceptable
B: Report production delayed because of illness	Alternate arrangement not done	Low	Have alternate persons	Minimal	Multi skill training and team working

7.5 Simple rating methods

There are number of methods for rating the problems as well as for the remedies. Two simple rating method techniques are described in Mycoted that are used for the initial sorting of large numbers of ideas. The first one is the rating adopted by Moore as simple/hard/difficult and the second one is "✔ ?W" approach. They are very useful for quick initial screening, but both the approaches have the disadvantage that they may lead to a rather superficial and potentially unreliable sorting of ideas and may ignore other criteria.

The "✔ ?W" approach can be more realistic in that "✔ " is only used for cases where implementation is relatively obvious, the other two categories reflect intuitive appeal, rather than objective evaluation.

(1) Simple/hard/difficult: Moore rated the ideas as simple, hard, and difficult in 1962. The creativity groups are expected to work through

their list of ideas and make judgments as to the priority rating they feel is appropriate, each idea should be marked as follows:

- *Simple*: Feasible with a minimum of time and money.

- *Hard*: Feasible, but will be more expensive.

- *Difficult*: Feasible but much more expensive.

(2) ✔ ?W (Okay-what/why-weirdo): The ✔ ?W method is comparable to the simple/hard/difficult method mentioned above, but aimed at cases where the creativity team will do their own evaluation so the criteria are much closer to the creative process.

✔ : Ideas those are feasible as they stand, they are generally ideas you would be happy to show to the client.

?: Ideas those are not feasible as they stand but have potential with more thought or research, or in the future, or under special circumstances.

W: Stands for 'weirdoes' – ideas that are bizarre and totally unfeasible as they stand, but have the potential as 'De Bono intermediate impossibilities' for further idea generation like using 'crazy' ideas.

One can develop his own rating methods by depending on the importance given for the problem and the number of ideas got. It might be "good, bad, and ugly", "bronze, silver, gold and platinum" or ranking in a scale of 1–5 or 1–10. The ranking should be easy for the people to understand and take decision.

7.6   Stakeholder analysis

Stakeholder analysis developed by Mason and Mitroff in 1981 looks at how groups of people might affect the outcomes of a proposal by the way they react. Stakeholder analysis is the technique used to identify the key people who have to be won over. To identify stakeholders, the checklists are prepared. Examples are as follows:

- Who are the sources of reaction or discontent to what is going on?

- Who have relevant positional responsibility?

- Whom do others regard as 'important' actors'?

- Who participate in activities?

- Who shape or influence opinions about the issues involved?

- Who fall in demographic groups affected by the problem?

- Who have clear roles in the situation (e.g., customer/friend/adviser)?

- Who are in areas adjacent to the situation?

One can make use of a matrix like the one below by which stakeholders can be plotted and categorized both by the chance of their affecting the situation, and by the scale of impact they would have if they did. Should any quadrant in the matrix appear empty, check that you have really included everyone, or plot the scale of the stakeholders' influence (high or low) against whether they would support or oppose your project.

	Impact unlikely	Impact likely
Impact, if it occurred, would be high	Chairman of the board Chief accountant	My manager Key customer
Impact, if it occurred, would be low	Reprographics Department	My secretary

Listing any assumptions that stakeholders are making could prove helpful e.g., using assumption surfacing, carefully assessing the list, especially in relation to the stakeholder for whom they have been derived. Ask yourself, does this actor have any special power in the situation; and if so, are there any of his or her assumptions that could have a considerable effect on your project? How this stakeholder could be influenced to change their point or course of action? A careful understanding of the roles and influences of stakeholders can help in taking correct decision.

The benefits of using a stakeholder-based approach are as follows:

(1) You can use the opinions of the most powerful stakeholders to shape your projects at an early stage. Not only does this make it more likely that they will support you, their input can also improve the quality of your project.

(2) Gaining support from powerful stakeholders can help you to win more resources – this makes it more likely that your projects will be successful.

(3) By communicating with stakeholders early and frequently, you can ensure that they fully understand what you are doing and understand the benefits of the project – this means that they can support you actively when necessary.

(4) You can anticipate what people's reaction to your project may be, and build into your plan the actions which will win people's support.

The first step in stakeholder analysis is to identify who your stakeholders are. The next step is to work out their power, influence, and interest, so you know who you should focus on. Each stakeholder has certain rights and responsibilities with harms and benefits, as a result of their actions. The final step is to develop a good understanding of the most important stakeholders so that you know how they are likely to respond, and so that you can work out how to win their support – you can record this analysis on a stakeholder map. After you have used this tool and created a stakeholder map, you can use the stakeholder planning tool to plan how you will communicate with each stakeholder.

In any transaction, you can identify the stakeholders as lenders or borrowers. Identify the rights and responsibilities of the stake holders, what harms they can make, and what benefits they are going to reap while making analysis.

Take example of a textile mill. The stakeholders are shareholders, top management, employees, suppliers, customers, community, and the local government. Balancing the needs including egos of all stakeholders is essential while finding remedy for any problem. Any imbalance can lead to resistance and the remedy cannot be implemented. Let us look at few cases:

(1)   On a particular year the mill made profit and management wanted to give 12% bonus, whereas the union was expecting 10% only. When the management told that they want to give 12%, the union leader objected. He said "You cannot declare bonus of 12%. What is my prestige? You declare only 8.33% and I will demand 12%. I will call for a strike and you compromise for 10%. Do not give more than 10% as next year you cannot give, but I will demand. If you have extra money, give it in some other form but not as bonus". There was logic in the union leader's statement and management agreed for that. There was a need for the union leader to maintain his image as leader of workers who struggled and got 10% bonus. If management had declared 12%, then the leader would not have been regarded by the workers.

(2)   A waste spinning mill had old machines and the layout was congested. There used to be frequent fire accidents. The mill had open land around the factory building, and hence the management decided to expand the building and relocate the machines in a spacious layout. There was a small idol of Lord Hanuman outside the building and workers used to worshiping it before coming to work. The management constructed a good temple for Hanuman and declared the date for relocation. As this point was not discussed with union leaders before taking a decision, they opposed the shifting of idol, but insisted the shifting of factory building itself, or expand the building without touching Lord Hanuman. There were frivolous discussions, and the workers were not ready to allow the shifting of the idol even at the cost of their lives. Finally, the management agreed to their suggestion and expanded the

building in such a way that Lord Hanuman came inside the factory. As the building was expanded, all old machines were removed and new machines were installed. Cabling was done in a secured way, good ventilation was provided and finally, the fire accidents could be prevented. There is not a single fire accident in that plant since last 20 years. The management claimed that the fires were prevented because of the new well-designed building and the technology adopted; whereas the workers say that it because of Lord Hanuman, who is now inside the factory.

(3) In a weaving factory at South Gujarat, the chief executive wanted to improve the systems. The empty warper beams were lying on the floor and it was a hindrance for free movement and hence, he ordered for racks. The managing director who was not consulted before ordering the racks did not approve the purchase. The supplier was demanding the payment. Managing director insisted to return the materials and was ready to bear the transport expenses and some penalty but did not allow those racks to be installed. Finally, the chief executive had to resign and leave. After two years, the managing director ordered for the racks of the same design and installed them.

7.7   Strategic assumption surfacing and testing

Strategic assumption surfacing and testing (SAST) is a method for integrating world views developed in an organizational context. It is a method for approaching ill-structured problems that is based on the premise that we all live our lives according to the assumptions we make about ourselves and our world. An ill-structured problem is one for which various strategies for providing a possible solution rest on assumptions that are in sharp conflict with one another. To cope better, we need to surface those assumptions and challenge them. New assumptions then become springboards to effective change. This concept was developed by Mason and Mitroff in 1981.

The purposes of SAST method are as follows:

- To help surface for explicit examination of the underlying assumptions that analysts often unconsciously bring with them to a problem situation.

- To compare and to evaluate systematically the assumptions of different analysts.

- To examine the relationship between underlying assumptions and the resulting policies that are derived and dependent upon them.

- To attempt to formulate new, novel, and originally unforeseen policies based on previously unforeseen assumptions.

This method assists the participants to understand a problematic situation and explore strategies for dealing with it. As the name indicates, its central element is to bring to the surface the assumptions that underlie people's preferred approaches to an issue, and to challenge them. This act of challenging, sometimes, may result in a particular strategy being discarded and participants adopting a competing one. On other occasions, however, integration occurs through the synthesis of previously inconsistent assumptions, resulting in a new strategy that accommodates the differences between those held initially, and which is stronger than the components from which it arises.

Four stages in the method include the following:

(1) Assumption specification

(2) Dialectic phase

(3) Assumption integration phase

(4) Composite strategy creation

Four principles are explained in Chapter 4 of "Dialogue methods for understanding particular aspects of a problem: integrating visions, world views, interests and values" of the book by epress.anu, underlie the strategic assumption surfacing and testing method: it is adversarial, participative, integrative, and 'managerial mind supporting'.

- *Adversarial* – Based on the belief that judgments about ill-structured problems are best made after consideration of opposing perspectives.

- *Participative* – It seeks to involve different groupings and levels in an organization, because the knowledge and resources needed to solve complex problems and implement solutions will be distributed around a number of individuals and groups in the organization.

- *Integrative* – On the assumption that the differences thrown up by the adversarial and participative processes must eventually be brought together again in a higher order synthesis so that an action plan can be produced.

- *Managerial mind supporting* – Believing that managers exposed to different assumptions will possess a deeper understanding of an organization, its policies, and problems.

The four steps used in the method are as follows:

(1) *Group formation*: Gathering as many people as possible who are involved in and affected by a situation, and splitting them into small groups according to their views on key issues. It is important to minimize the conflicts within each group and to maximize the differences between groups. The orientation to the problem held by

each group should be directly opposed by at least one other group. If there is no opposition for the views given by any team, then the chances of getting the real root is less. The people may get biased.

(2) *Assumption surfacing and rating*: Identifying the preferred strategy or position that each group is adopting, then revealing and quantifying (if possible) the assumptions on which it is based. Techniques used include stakeholder analysis, assumption specification, and assumption rating.

(3) *Intra-group and inter-group dialectical debate*: Each group developing the case for its position and then discussing them all in a single large group. The process is dialectical 'if it examines a situation completely and logically from two different points of view'. A key analytical question that facilitates dialogue in this stage is – What assumptions of the other groups does each group finds the most troubling.

(4) *Final synthesis*: Achieving an accommodation among participants to find a practical way forward. Discussion of key assumptions leads them to be modified and a new strategy to be developed, based on the modified and agreed-on assumptions. This is the process through which the visions and world views of the participants become integrated. If agreement cannot be reached (if synthesis is not possible), participants might agree on a program of research or other action to further clarify assumptions and/or to try out a particular strategy and evaluate it. Knowledge gained from those steps can shed further light on the conflicting assumptions and facilitating subsequent synthesis of positions.

This approach to strategic planning has been contrasted by its originators Mason and Mitroff with the two dominant approaches namely, the 'expert' approach in which an organization establishes a planning unit to largely do the managers' work for them, and the 'devil's advocate' approach in which middle managers prepare and submit plans to senior managers for cross-examination.

The strategic assumption surfacing and testing method was developed as a contribution to strategic planning and not as a problem tool. It has great potential where conflicting views on the nature of a problem and what to do about it are held, where the proponents are willing to work in groups to explore these issues and are open to hearing and understanding others' views, with the aim of finding accommodation between the originally conflicting positions. Being willing to reveal, explore, and expose to criticism of the assumptions that one brings to the process, and a concomitant willingness to challenge others' assumptions, are essential to the achievement of synthesis.

Strategic assumption testing explained in Mycoted examines other people's opinions and assumptions to ensure that they are consistent. It has a number of steps which are as follows:

Stakeholder identification: List those involved as stakeholders. If in groups, each group should make its own list privately and then collate. Identification of stakeholders is very important as the success of an action depends mainly on the cooperation given by the stakeholders.

Identify factions: Identify the factions and if necessary, group stakeholders into factions like 'points of view' or 'interests'. There might be different sub groups or factions within a group of stakeholders. If we consider customer as stakeholder, the factions can be area-wise customers, product-wise customer, big/large customers, small customers, area wise agents, retailers, wholesalers, etc.

Group formation: Establish one or more working groups from the 'sub-groups'.

Assumption surfacing: In each sub-group, discuss each stakeholder's reasons (assumptions) and prioritize them. Try understanding from the stakeholder's point of view and not your assumption or experience. You can make role plays to understand the stakeholder's views.

Assumption testing: Members of the sub-groups debate if these assumptions were reversed and if it made no difference than should we ignore it. Testing is to be done from all angles, i.e. 360° in 3 dimensions.

Assumption ranking: Members of the sub-group rank their assumptions:

- Effect if the assumption occurred

- Possibility of it occurring

Results are exhibited as a 2 × 2 matrix of high/low potential versus likely/unlikely occurrence.

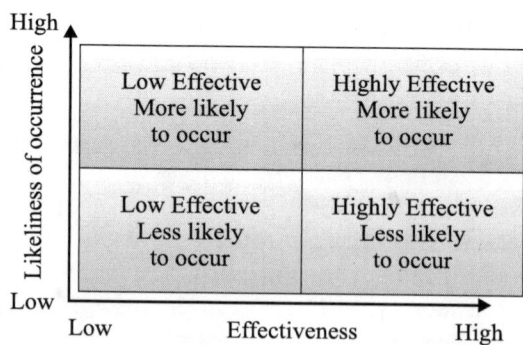

Figure 7.3 Effective occurrence matrix.

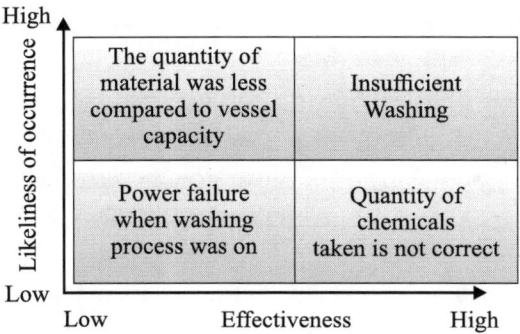

Figure 7.4 Case of low wash fastness of dyed yarn.

Action planning: Members of the sub groups analyze the 2×2 matrix and its possible consequences. They suggest the plans for tackling different situations.

Inter-group debate: Each sub group puts forward their matrix and plan, this generates an open debate. Issues are identified and fundamental assumptions are challenged. Ultimately, a common ground is sought.

7.8   Strategic choice approach

Taking decision under pressure is different from taking decision with a cool mind, when you can think, discuss with friends and well-wishers. When under pressure, there shall be no time to think. The strategic choice approach to planning under pressure has been developed by OR scientists as a means of facilitating communication among decision-makers with diverse perspectives, allegiances, and skills. Its function is to enable them to make sustained progress together in exploring the structure of complex decision problems, and in charting progress towards timely commitments to agreed actions. It offers a balanced set of communication tools which are primarily visual in form, yet are interlinked within a philosophy of planning that recognizes the challenges that decision-makers face in responding strategically to diverse sources of uncertainty, including those that call for a political or structural rather than an analytical response. The principles and leading tools of the approach are briefly introduced, with references to the growing range of applications in fields of collaborative planning ranging from local community action to national environmental policy

The strategic choice approach by John Friend and Allen Hickling appeared in 1987 as a means of Planning under Pressure. The approach has been gathering support from decision makers, while also becoming widely taught in management, planning, and policy schools. There are four essential steps suggested that are as follows:

(1) Focuses on decisions to be made in a particular planning situation, whatever their timescale and whatever their substance.

(2) Highlights the subtle judgments involved in agreeing how to handle the uncertainties which surround the decision to be addressed – whether these be technical, political, or procedural.

(3) The approach is an incremental one, rather than one which looks towards an end product of a comprehensive strategy at some future point in time. This principle is expressed through a framework known as a "commitment package". In this, an explicit balance is agreed between decisions to be made now and those to be left open until specified time horizons in the future.

(4) The approach is interactive, in the sense that it is designed not for use by experts in a backroom setting, but as a framework for communication and collaboration between people with different backgrounds and skills.

There are three key elements of analysis that are used in structuring problems and working towards decisions:

(a) The decision area

(b) The comparison area

(c) The uncertainty area divides into three broad categories:
   • Uncertainties to do with the working environment
   • Uncertainties to do with guiding values
   • Uncertainties to do with related choices

A repetitive technique used for complex problems and their sub-problems consists of four basic principles that are as follows:

• Shaping – It involves the identification of the problem areas.

• Designing – It recognizes what can be done by looking at possibilities and drawbacks.

• Comparing – It involves comparison of various ideas, evaluating the best possible way forward.

• Choosing – It involves choosing the best ideas for solving the problems. Compiling a plan of action and acknowledging any uncertainties.

Figure 7.5 Typical situation when under.

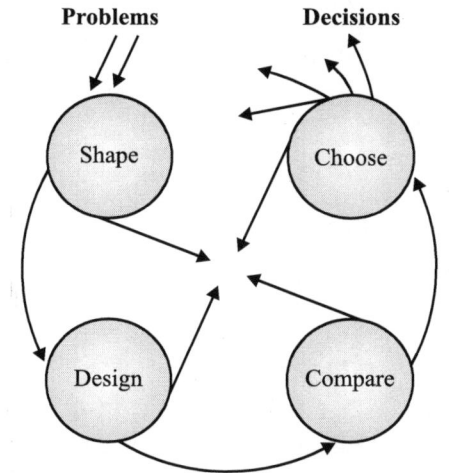

Figure 7.6 Four modes of strategic choices.

If we consider the remedies suggested for avoiding the shortage of skilled people, we may find that the real problem is not in the low salary paid by the company but the way in which the people are treated by their bosses. It is needed to educate the bosses to treat the people under them in an ethical way. But how to do it is a problem, as the bosses think that they know how to handle the situation and that is why they have been made bosses by the management. They neither are ready to listen to any one, nor would they read any journals. The ego is the biggest problem because of which they are not treating their subordinates well. They want to show that they are supreme and are always afraid that their subordinates might overtake them. Such people shall not like to sit and listen in a training class, if they are asked to attend. They feel it as an insult for them. They are not ready to believe that there can be a person better knowledgeable than them. They think what they learnt in their course of education and experience is the ultimate as they have struggled and learnt a lot which is not explained in any book and not taught in any school or college. Designing a programme for them should be in an intelligent way. They should be called as chief guests for the training programme organized for their subordinates, where the faculty explains the importance of treating subordinates in a decent way. In the class, case studies may be given with different permutations and combinations, and the pros and cons of various actions should be explained. As chief guest, the boss shall be listening to the faculty, and finally he would realize his mistake and shall try to change himself gradually.

7.9   Strategic management process

This is a six-stage process, run in-house usually by a strategic management group. It is supported by various consultants and accessible to external stakeholders. This method is normally used by public and voluntary organizations, and can also be used for textile and apparel industry. It has a number of processes that are as follows:

(1) Historical context – examination of previous trends, and the emergence of a future vision for the way ahead. Although we say textile and apparel are driven by unpredictable fashion, a careful study can give a clue as to which fashion is likely to dominate. The future predictions done by fashion experts are based on historical data combined with new developments in the technology and products.

(2) Situational assessment – blame free SWOT analysis of the present situation. Study the actual situation and find out why it is like this. Whether a trend seen is temporary or likely to remain for a long time. The recessions normally found are due to various reasons like excess production, high cost of products because of increase in cotton prices and increase in cost of living, increase in prices of some of the essentials like food, water, power, etc., analyse what can be reversed.

(3) Strategic issue agenda – It identifies the issues from points 1 and 2 above and acknowledges the relationships that exist between other points stated below.

(4) Strategic options – Define as many positive solutions to meet the SWOT analysis and future vision. Define strategies, and outline costs, feasibility, acceptability and effectiveness. You may consider shifting from cotton to polyester or from polyester to cotton, dress materials

to sarees or sarees to dress materials, labour intensive systems to automation or commodity products to niche products, etc.

(5) Feasibility Assessment – A selection of strategies is examined through stakeholder analysis and resource analysis. Identify which is more feasible to your stakeholders and what resources are easy for you to provide.

(6) Implementation – It is to evaluate the stakeholders' predictions, a series of evaluation programmes is devised.

Within each stage above, three basic steps are followed:

- Search – For ideas and information
- Synthesis – Observation of patterns, trends
- Selection – Determine priorities for action
- Within these three basic steps, four alternative criteria are used to assist by using the best technique.
- Quality
- Acceptance
- Innovation
- Preservation

The alternative which is most economical or maximum rewarding while acceptable to majority of stakeholders is selected.

The strategic management process is more than just a set of rules to follow. It is a philosophical approach to business. Upper management must think strategically first, then apply that thought to a process. The strategic management process is best implemented when everyone within the business understands the strategy. The five stages of the process given by Jim Clayton and Demand Media are goal-setting, analysis, strategy formation, strategy implementation, and strategy monitoring.

Goal-setting: The purpose of goal-setting is to clarify the vision for your business. This stage consists of identifying three key facets: First, define both short- and long-term objectives. Second, identify the process of how to accomplish your objective. Finally, customize the process for your staff; give each person a task with which he can succeed. Keep in mind that during this process your goals should be detailed, realistic, and match the values of your vision. Typically, the final step in this stage is to write a mission statement that briefly communicates your goals to both your shareholders and your staff.

Analysis: Analysis is a key stage because the information gained in this stage will shape the next two stages. In this stage, gather as much information and data as you can which is relevant to accomplish your vision. The focus of the analysis should be on understanding the needs of the business as a sustainable entity, its strategic direction, and identifying initiatives that will help your business grow. Examine any external or internal issues that can affect your goals and objectives. Make sure to identify both the strengths and weaknesses of your organization, as well as any threats and opportunities that may arise along the path.

Strategy formulation: The first step in forming a strategy is to review the information gleaned from completing the analysis. Determine what resources the business currently has, which can help reach the defined goals and objectives. Identify any areas in which the business needs to seek external resources. The issues facing the company should be prioritized by their importance to your success. Once prioritized, begin formulating the strategy. Because business and economic situations are fluid, it is critical in this stage to develop alternative approaches which target each step of the plan.

Strategy implementation: Successful strategy implementation is critical to the success of the business venture. This is the action stage of the strategic management process. If the overall strategy does not work with the current structure of the business, a new structure should be installed at the beginning of this stage. Everyone within the organization must be made clear of their responsibilities and duties, and how that fits in with the overall goal. Additionally, any resources or funding for the venture must be secured at this point. Once the funding is in place and the people are ready, execute the plan.

Evaluation and control: Strategy evaluation and control actions include performance measurements, consistent review of internal and external issues, and making corrective actions when necessary. Any successful evaluation of the strategy begins with defining the parameters to be measured. These parameters should mirror the goals set in Stage 1. Determine the progress by measuring the actual results versus the plan. Monitoring internal and external issues will also enable you to react to any substantial change in your business environment. If the strategy is not moving the company towards its goal, take corrective actions. If those actions are not successful, then repeat the strategic management process. Because internal and external issues are constantly evolving, any data gained in this stage should be retained to help with any future strategies.

7.10   Assess by five golden questions

The five golden questions developed by B. Purushothama in 1996 is a strong self-assessment tool, which can be used at any place. It helps in understanding whether we are moving in the correct direction or not. We need not give answer to anyone, but have to answer our consensus. If we can convince ourselves that we are moving in the correct direction, then we are bound to get the result. The questions are as follows:

The first question is "Whether we have a procedure?" For doing any work, there should be a procedure, i.e. a defined method of working that is established. The procedure may or may not be documented, but should be in practice. Every one should be working in that style without fail all the time. The procedures should have been established by judicial studies and logical thinking, and not just by someone's ideas. The procedure should have been tried, proved, and established before declaring and documenting it as a procedure. First, verify whether there is an established procedure for the work you intend to do or for the work you are already doing. If procedure is not there, your first work should be establishing a procedure.

When you say you have found a remedy for a problem, establish the procedure for people to follow so that the problem does not recur; or if it recurs, people take immediate action without waiting for their bosses to give clearance. The procedure should be available to the people on the job and all should be educated to refer and work as per the procedure.

We may have a procedure, but only having is not adequate. It should be appropriate for the situation and suitable for the activity in hand. The procedure should be clear. It should guide in each step of the activity and address the objectives. It should be suitable to achieve the objective in time with least expenses or efforts. Hence, there is a need to periodically review the procedures to ensure their suitability considering the changing situation and environment. The procedures should be tailor made for the situation and not to be copied from others. The procedure should be appropriate for the culture, technology, infrastructure, environment, and the expectations of the customers. There might be different ways and means to do a work or to achieve the objectives.

If you analyse problems in textile and garment industries, you can see that invariably the people are following the procedures given by their bosses but are not getting the results as anticipated by the management. It means the procedure given by the boss is not suitable to achieve the goal given by management. Therefore, only having a procedure cannot give results.

Writing a procedure is very easy as compared to its implementation. It is more complicated when it has to be implemented organization wide. Normally, implementation means organization wide implementation without any excuse. One can make use of linking exercises in order to study the effect of implementation at one place and on the activities at other place.

In a number of textile mills we can see nicely documented procedures, often written by well-experienced consultants. The men on shop floor including the heads of the departments, the functional heads, and even the chief executive and managing directors have no time to read the procedures documented. The procedures are kept in lock, as it is needed to be shown to auditors when they come for compliance audit. They are never given to people down the line to read and understand. There are no systems of reading the procedures, discussing it in a group of people, and implementing that procedure. In such cases, the procedures are not said to be implemented and we cannot expect results to be achieved.

Any procedure as stated earlier should address the objectives. If the procedure is evaluated and proved as the best to achieve the objectives; and if we were able to implement it in real sense, then we must get the results as anticipated. If the results are not obtained, it means either our procedure was not suitable or we did not implement it as required. A careful analysis of the situation can reveal the actual reason for not achieving the required goal.

Monitoring and measuring the activity is very important in order to achieve the results. We need to learn measuring everything, and then only we will be able to say whether we got the result as anticipated. If we do not know how to measure, then we will not be able to monitor it.

We might achieve the results as anticipated or committed. It does not mean that we are efficient. Our success in a competitive environment depends mainly on the performance of our competitors. In a running race, in order to become a winner, the speed at which I am running is less important than the distance I am keeping ahead of my nearest competitor. Although we achieved the results as anticipated, we cannot be happy for that. If our competitors are doing a better job, we are certainly going to lose. Therefore, it is always necessary to keep a watch on the competition and develop our systems to achieve better results. We need to benchmark the best in each criterion, and work towards meeting and overtaking. We need to compare ourselves with those who are stronger in various fields.

The above questions can help us in assessing our efforts to solve the problems.

8.1 Dealing with the change

Identifying a problem and finding a solution might be an easy task as compared to implementing what was decided. One has to face a lot of challenges while implementation. We shall be bringing a change while implementing a remedy for a problem. We might bring change in people, change in technology, change in thinking, change in policy, change in procedures and systems, change in environment; which all tend to affect the equilibrium. Modification in the way of performing certain jobs, change in rules and procedures, adoption of a new technology, changes in organizational structure, etc., affect the internal equilibrium; whereas change in market situation, government rules and regulation, political scene, economic scenes, etc., affect the external equilibrium. One has to deal with the change to reduce the tension by understanding and altering the forces and making people participative and committed.

Change management entails thoughtful planning and sensitive implementation, and above all, consultation with, and involvement of the people affected by the changes. If one forces change on people, normally problems arise. Change must be realistic, achievable, and measurable. These aspects are especially relevant to managing personal change. Before starting organizational change, one needs to ask self: What do we want to achieve with this change; why and how will we know that the change has been achieved? Who is affected by this change, and how will they react to it? How much of this change can we achieve ourselves, and what parts of the change do we need help with? These aspects also relate strongly to the management of personal as well as organizational changes.

There are effective ways of overcoming problems set by change. One adapts to change by seeing what is changing and how it is changing. One estimates what is likely to happen, plans ahead, and acts accordingly. Planning includes evaluating alternative strategies to find the best strategy. It is this best strategy which then becomes part of our forward plan.

In order to bring change effectively, one needs to understand the following:

- The personal views, values, and experiences of individuals.

- The culture of the organization.

- The beliefs, those which have deep roots.

- Possibilities for resistance to change.

- The possible reasons for resistance to change.

- Reasons for going along.

- How change affects the individual.

- How to help people to move smooth through the road to change.

- Typical patterns of behavior of individuals and group.

- Dealing with setbacks, slowdowns, and uncertainty.

Various steps in managing change include the following:

- Assessing the need for change

- Designing the plan for change

- Coaching those who will lead others through the transition to change

- Helping others adapt to change

- Dealing with resistance to change

8.2   Resistance to change

Change is taking place and things are changing more rapidly. We now live in a time of change; we live in a time of accelerating change. This is how our life today differs from the life of those living a few years ago. There is the impact of technology, laser applications, electronic calculators, desk computers, computer-based automatic operations and control, microprocessors, and much more. While on the one hand our spaceships have explored our planetary system and we have landed on the moon, on the other hand some country uses psychiatric hospitals in an attempt to break the resistance of those who stand up for individual freedom by disagreeing with the dictates of the state. Some established norms have been questioned. People's behavior is being affected, and there has been an increase in permissiveness (promiscuity), delinquency, and crime; and those who are involved mostly are of much younger age.

The old house caught fire, we need to change. Some are reluctant to change

Figure 8.1 Resistance to change.

Most people have some discontent regarding the current order and would like something to be altered. But those who benefit most from the current situation are those in power. They have the most influence, the most to lose, and the strongest motive to keep things pretty much the same way as they are. While initiating a new system, normally some resistance is seen. Many people in the organization, including a proportion of the management team or top people, oppose the very idea of any real change in the status quo. The reasons might be self interest, misunderstanding, and lack of trust or fear of unknown, different assessment or bitter experiences of past, low tolerance for change, the power of the status quo, etc. The people somehow do not readily accept a change and like to continue in the same situation or system which they are in practice for a long time. When the resistance to change is examined, we find that we are dealing with two changes; a technological change, and a social consequence of the technological change.

When people are confronted with the need or opportunity to change, especially when it is 'enforced', as they see it, by the organization, they can become emotional. So can the managers succeed who try to manage the change? Diffusing the emotional feelings, taking a step back, encouraging objectivity, are the important points to enable sensible and constructive dialogue.

Strong resistance to change is often rooted in deeply conditioned or historically reinforced feelings. Patience and tolerance are required to help people in these situations to see things differently. Planning, implementing, and managing change in a fast-changing environment is increasingly the situation in which most organizations now work. This is true for textile world also.

Some of the methods used for overcoming a resistance are as follows:

8.2.1 Participation and involvement

Success depends to a large extent on the commitment of employees towards the organization's aims, and on cooperation between themselves. This means identifying with the company. What benefits the company most is needed to be seen and felt by the employees, as it also benefits them; and the employees could be expected to work to targets that are set annually. The ways in which the targets are agreed, progress is monitored, and an individual is rewarded is crucial for the success of the organization or enterprise.

The initiators listen to the people who are involved in the change and use their advice. The participation leads to commitment and not merely compliance of a job; however, it might take more time. Members participate both in planning and execution of change and sufficient time is provided to evaluate the merits of the change in relation to the threat to their habits, status, and belief.

Normally resistance is seen while implementing new systems of working like going in for a documented quality management system; say ISO 9000. The resistance mainly comes from the heads of the departments and their direct reports, whereas the workers eagerly come forward to implement new systems. Making teams comprising of heads of the departments and senior staff helps in implementing the systems.

8.2.2 Facilitation and support

Strong resistance to change is often rooted in deeply conditioned or historically reinforced feelings. This happens when people are reallocated in jobs due to changes in technology or shifting of venue, introducing new concepts of management, etc. Patience and tolerance is required to help people in these situations to see things differently. Although people are willing to accept the change, initial failures make them feel unsecure with the new system and hence they resist. Training is provided in new skills, and sufficient time is given to employees for implementing a change. Facilitation and support are most helpful when fear and anxiety lie at the heart of resistance. The change can be implemented fully when the people develop confidence.

It is suggested not to raise more nonconformity in the initial audits when a new system is being audited, as people may feel insecure with the new system. Identify simple areas where they can take action first and gradually improve your standard of auditing.

8.2.3 Negotiation and agreement

Incentives are offered to active or potential resistors, so that they can agree for bringing the change. It is particularly appropriate when it is clear that someone is going to lose out as a result of a change and his power to resist is significant. It is suggested not to sell change to people as a way of accelerating "agreement" and implementation. "Selling" change to people is not a sustainable strategy for success. Instead, change needs to be understood and managed in a way that people can cope effectively with it. Change can be unsettling, so the manager logically needs to be a settling influence.

8.2.4 Manipulation and cooperation

Co-opting an individual by giving him a desirable role in the design or implementation of a change is a form of manipulation. This involves selective use of information and conscious structuring of events that could be used to deal with resistance. It is suggested to check that people affected by the change agree with, or at least understand, the need for change, and have a chance to decide how the change will be managed, and to be involved in the planning and implementation of the change. Face-to-face communication to handle sensitive aspects with people is suggested if you want to manage an organizational change. Email and written notices are extremely weak at conveying and developing understanding.

8.2.5 Explicit and implicit coercion

People are forced to accept a change by explicitly or implicitly threatening them, or by actually firing or transferring them. This is a risky process as people strongly resent forced change. This step is taken when the majority is convinced and has accepted the change and is willing to go with the management, and very few are opposing.

For organizational change that entails new actions, objectives, and processes for a group or team of people, using workshops to achieve understanding, involvement, plans, measurable aims, actions, and commitment are suggested. These are suggested even to very tough changes like making people redundant, closures, and integrating merged or acquired organizations. Bad news needs even more careful management than routine change. Hiding behind memos and middle managers will make matters worse. Consulting with people and helping them to understand does not weaken your position, although it strengthens it. Leaders who fail to consult and involve their people in managing bad news are perceived as weak and lacking in integrity. Treat people with humanity and respect, and they will reciprocate.

8.3 Chaturopaaya

Four steps for bringing a change is explained in Indian Subhashitas as "*Saama*" (साम), "*Daama*" (दाम), "*Bhedha*" (भेद) and "*Danda*" (दंड). *Saama*, the first step is to explain and convince, i.e. the process of pacifying. The second step *Daama* is to explain the consequences of not accepting a change in monitory terms or the advantage one gets by adopting the change. This also involves paying something in order to get the work done. *Bhedha* is the principle of divide and rule, i.e. to differentiate or separate the people who are accepting a change from those who are not ready to change. *Danda* is to punish, the last resort for bringing a change. These are called as *"Chaturopaaya,"* i.e. four ideas ['*Chatura*' (चतुर) = Four, '*Upaaya*' (उपाय) = Idea]. The above four '*Upaayas*' are the base for all the problem-solving techniques practiced world over.

The chapter 7.5.19 of *Srimad Bhagavatam* written by Maharshi Vyasa includes the Chaturopaya. The teachers Sanda and Amarka thought that Prahlada Maharaja was sufficiently educated in the diplomatic affairs of pacifying public leaders, appeasing them by giving them lucrative posts, dividing and ruling over them, and punishing them in cases of disobedience. Acharya Chanakya explained the same techniques to Chandragupta Maurya, so that he could become a king and later emperor by using these techniques.

It is essential for a student who is going to be a ruler or a king to learn the four diplomatic principles. There is always rivalry between a king and his citizens. Therefore, when a citizen agitates the public against the king, the duty of the king is to call him and try to pacify him with sweet words, saying, "You are very important in the state. Why should you disturb the public with some new cause for agitation?" If the citizen is not pacified the king should then offer him some lucrative post as a governor or minister (any post that draws a high salary), so that he may be agreeable. If the enemy still goes on agitating the public, the king should try to create dissension in the enemy's camp, but if he still continues, the king should employ severe punishment by putting him in jail or placing him before a firing squad. The teachers appointed by Hiranyakashipu taught Prahlada Maharaja to be a diplomat so that he could rule over the citizens very nicely.

The first step in bringing a change is always educating the concerned about the problem, its pros and cons, the need to change, and the likely effects of the change, how it is going to benefit self and the society, the role of each one in bringing the change, etc. *Sama,* the first step, is explaining to people by word, common sense, and logical explanations. *Daama*: the quintessential carrot. This second option is explaining along with giving an incentive. *Bheda*: the third option is used if *sama* and *daamma* doesn't work. This is the art of selective discrimination and differentiation. *Danda*: if none of the above methods work, then the last resort is to use force or the stick.

This is a political methodology to approach a given situation. Start with conciliation or gentle persuasion (*Sāma*). If it does not help, then offer money/material wealth, i.e. (*Dāna or Daama*). If that still does not change the status quo, use threat or cause dissension (*Bheda*). Use punishment or violence (*Danda*) to resolve the situation where the previous three fail. Use of illusions or deceit [*Māya* (माया)], deliberately ignoring people [*Upeksha* (उपेक्षा)], and use of jugglery [*Indrajāla* (इन्द्रजाल)] are also suggested to resolve any situation.

One should note that *Danda* should be the last resort, and it should not be done first. If you start doing it first, people shall leave you and your organization would not survive; forget about bringing any change.

The "mind tools" explains the four essential steps for bringing a change which emphasizes on educating and bringing awareness among the people. The steps are involved are as follows:

(1) Ensure that everyone understands that why change is necessary. If people are dissatisfied with the way the things are, they will be more likely to welcome change. So explain the demerits of present system.

(2) Show the concerned people how things will be better in the future. What advantages they are going to get.

(3) Ensure that people understand the plan. Do not just assume that people have understood as you gave a nice presentation and there was no query.

(4) Try to ensure that there can be no way of going back to previous ways of doing things: ensure that only new forms are available, that computer systems reflect the new way of working, and that procedures work smoother under the new system than the old ones.

Allan Chapman cautions not to sell change to people as a way of accelerating agreement and implementation. "Selling" change to people is not a sustainable strategy for success. Change needs to be understood and managed in a way that people can cope effectively with it. Change can be unsettling, so the manager logically needs to be a settling influence. Mehrabian suggests the use of face-to-face communications to handle sensitive aspects of organizational change management. He suggests encouraging managers to communicate face-to-face with their people, if they are working for managing an organizational change. Email and written notices are extremely weak at conveying and developing understanding.

Hrebiniak, author of *Making Strategy Work*, stresses that to change culture, you should focus on four of the factors and conditions that affect it that are as follows:

(1) Structure and process: Large retail stores are seeking to achieve decentralized operations and create a culture of decision-making autonomy so that they can get close to customers and local tastes; they might ask corporate and regional managers to leave stores alone and allow store managers to do their own thing. Interference with the stores is likely to decrease if the managers are asked to stay away and

let local decisions and actions prevail. But what happens when the next major problem arises? Corporate or regional managers swoop down on the stores, bringing centralized solutions. As an alternative, they should consider changing the structure. Increasing the span of control for corporate or regional managers would militate against involvement in the stores. Large spans foster decentralization and autonomy at lower levels by making it more difficult to actively meddle in a larger number of stores' strategy and operations. Behavioral change of top managers can foster behavioral and cultural change in the stores.

(2) People: They bring in fresh blood and thinking. Rotate managers with different views of competitive conditions or operations. Supply different, needed skills or capabilities from the outside. New people, ideas, and strategies can lead to behavioral and performance changes that, in turn, can affect new ways of thinking and culture change. This should be done while keeping the people in confidence or else, there are chances of losing good people we already have. There are few textile mills which have brought marketing people from different disciplines like automobile industry, home appliances, etc., for their retail business and they find it good. A man from a different industry or culture can give a number of new ideas to solve the chronicle problems because of his different angle of looking into the problem.

Whenever an organization imposes new things on people, there will be difficulties. Participation, involvement, and open, early, full communication are the important factors. Workshop is a very useful process in order to develop collective understanding, approaches, policies, methods, systems, ideas, etc. Staff surveys are helpful ways to repair damage and mistrust among staff; provided you allow people to complete them anonymously, and you publish and act on the findings.

Managers are crucial to the change process and hence management training, empathy, and facilitative capability are priority areas. They must enable and facilitate, not merely convey and implement policy from above which does not work.

You cannot impose change. People and teams need to be empowered to find their own solutions and responses with facilitation from managers. Management and leadership style and behavior are more important than clever process and policy. Employees need to be able to trust the organization. The leader must agree and work with these ideas, or else change is likely to be very painful and the best people will be lost in the process.

(3) Incentives: Randy Tobias once remarked that the culture of the old AT&T rewarded was "getting older." The culture, over time, became

stifling and bureaucratic. Appeals to managers to change and team-building exercises didn't work. But CEO Tobias and others after him changed incentives to reward performance, which was not getting older. New people were attracted by the new incentives and the opportunities presented, and the culture began to change. The same emphasis on incentives can be seen over the years in other similar companies. Incentives affect behavior and performance and attract new resources and capabilities, which can lead to culture change. This can very well be applied to textile and garment industries.

(4) Changing and enforcing control: It's important for companies to get feedback, evaluate performance, and take remedial action. Emphasis should be on concentrating strategy-implementation activities to achieve desired results. It is important to learn from performance, including mistakes, and use the lessons learned to change incentives, resources, people, methods and processes, and other factors to foster strategic and operating goals. It is also necessary to hold managers accountable for performance results. These actions or emphases will help to shape new behaviors, task interactions, and ways of thinking that will create or define a culture of learning and achievement.

8.4   Different models for bringing change

Allan Chapman in his article, "Change Management in "Business Balls," explains that change management entails thoughtful planning and sensitive implementation, and above all, consultation with and involvement of the people affected by the changes. If you force change on people, normally problems arise. Change must be realistic, achievable and measurable. These aspects are especially relevant for managing personal change. Before starting organizational change, ask yourself: What do we want to achieve with this change, why, and how will we know that the change has been achieved? Who is affected by this change, and how will they react to it? How much of this change can we achieve ourselves, and what parts of the change do we need help with? These aspects also relate strongly to the management of personal as well as organizational change.

Different models are available for organizing a change. Most popular among them are Unfreeze-Change-Refreeze, Change within, Learning and Power Sharing, Team Management and Human relation approach. Unfreezing is making people to recognize the need for a change. Once the need is felt, new methods and guidelines for change are introduced and applied. Refreezing stage provides required reinforcement to ensure that new behavior patterns are adopted permanently. The driving forces and restraining forces for a change are to be understood and balanced to have equilibrium. Internal distribution of power, internal mobilization of energy, and internal communication are very important factors in achieving a change. To bring a change within, one has to concentrate on correspondence between internal and external reality, goals, values, skills, and strategy. Learning helps a man to change himself, which need not be limited to classroom. Learning takes place at all levels of life and working. Unless the seniors or leaders set examples, juniors or followers cannot learn and implement new patterns. Along with learning, there should be an enhancement of power as successful change results from shared power and not from unilateral or delegated approaches. An approach of team management with high concern for people as well as production helps in achieving the change. This concentrates on team training, training inside the workplace for better integration between functional teams, arranging goal setting sessions, implementing plans, and finally stabilizing. In the human relation approach, managers are given training to understand human problems, diagnose the need, attitude and feelings of the people, and their capabilities, and to restructure the activities to implement the change successfully.

Different approaches are adopted in managing a change. Few examples are as follows:

(1) Information: Providing sufficient information helps in motivating an individual to accept the change, although that itself is not a motivational factor. Incomplete information or not providing information may lead to resistance.

(2) Individual counseling and therapy: This approach helps in achieving change at individual level by creating new insight, and is deeper compared to providing just information. One to one discussion and clearing the doubts helps in winning the other person.

(3) Influencing the peer group: It is observed that individual behavior is considerably influenced by the peers. The change process initiated in such a group is likely to be self-energizing and self-reinforcing. One need to identify the real peer and not the one designated as leader or peer.

(4) Sensitivity training: This training helps people to overcome their ego, understand the sentiments of others, cooperate for the common cause, and to achieve the required change. It helps one to realize that he is a human being and the others with him are also human beings.

(5) Feedback: Getting feedback and analyzing helps in correcting the situation, and builds confidence among the people working. The results need to be discussed in open, so that all involved can participate.

(6) Group therapy: In this approach, it is assumed that organizational conflicts are the result of individual characteristics and the approach consists of individual therapy and social psychology.

(7) Systematic change: This is highly powerful approach in changing organizations, which necessitate direct manipulation of organizational variables. The analysis is made of all the systems, and changes needed are discussed and decided, and the roles and responsibilities are redefined. In this approach, attempts are made to change the entire hierarchical distribution of the decision-making power in the entire organization or in the family.

(8) Technological approach: Efficiency, quality, and cost are the major considerations in a technological approach. Use of new technological methods that result in major changes in the authority and responsibility relations because of rampant mechanization, automation, and on-line controls, demands multi-skill jobs and multilevel organization structure.

(9) Value centered approach: This approach considers human motivation and personal growth by changing of values and norms prevalent in the organization or society. Total change is achieved by training and educating all from top to bottom, and atmosphere is created so that members of the organization, society, or family work with each other with mutual confidence and trust.

(10) Structural Approach: This approach employs the patterns of change in authority. The changes are initiated by top and are well planned. It may be an alteration of authority by decentralization and the establishment of different profit centers or an emphasis on formal structure, controls, and workflows.

There is no single simple formula for bringing a change. The guiding principles for managing a change as given by Hutton which are key success factors are as follows:

(a) If you do your homework, most of the issues that need to be dealt with are already addressed in some way in your plan. We fail mainly because our homework is not sufficient.

(b) You are not the first – many others have already done what you are attempting. Study the history and analyze.

(c) You can draw upon the practical experience of outsiders and other organizations to help you.

(d) You already have access to the most important sources of information to help you figure out what to do – your people and your organization.

There cannot be any better source of information other than your people and your organization.

(e) In working with individuals to win their commitment to change, start with their needs and ambitions – listen, don't preach.

(f) Work on building the commitment of the top key persons, and never stop reinforcing this commitment.

(g) Make the change process a team effort and, thus, ensure that everyone involved has the opportunity to take ownership of the process.

(h) Build partnerships which include all the key stakeholders; those who have the authority, the resources, and the expertise.

(i) In support of the transition, set priorities and focus your efforts where they can be effective. Work with enthusiasts who will lead the way and don't waste time trying to convert those who are lost causes.

(j) Strive for few small tangible early successes, and make the most of these through recognition and publicity. Do not keep a very big target in the first step itself.

(k) Ensure that those who are affected by the changes are involved in planning their own journey.

(l) Strive to act as a role model for others; do not be a preacher or dictator.

(m) Provide information to those affected about the need for change, the means, and also the effect of change on people.

(n) Listen and offer empathy for the stresses people undergo during change.

(o) Celebrate progress and make it fun.

By identifying the problem, analyzing the reasons, devising remedy, reinforcing the new system by carefully overcoming the resistance, change in the situation can be bought, which is beneficial. It can help you to win, and you can be a winner.

8.5 Ten principles for managing a change by Dr. Joel. R. DeLuca

DeLuca in his book "Overcoming Resistance to Change" explains ten principles for managing a change as follows:

(1) Don't use the model that made you successful. By using the same model repeatedly, we give an opportunity for opponents to develop a counter

strategy. Moreover we cannot bring a change, but maintain the same status quo by adopting the model repeatedly. *"Strategy Implementation is not Organizational Change"*. Few mills have developed the habit of offering something lucrative to the persons likely to resist. The resistors have learnt this and they accept the offering, but send someone else to create problem.

(2) Start at the end rather than at the beginning. The moment one activity is about to end, we should have started the next activity. *"Change is more of a circular than a linear, sequential process"*.

(3) Treat resistance as the organization's immune system. People who face resistance shall get immune to it. The one who tries to avoid resistance shall succumb easily. *"Don't resist resistance"*.

(4) Treat "head" or intellectual resistance as an issue of perspective. Intellectual resistance forces you to find alternatives and become determined in the work. It gives you immense strength. *"You have to see something new before you can do something new"*.

(5) Treat "heart" or emotional resistance as an issue of self esteem. *"Saving face is critical issue in making change happen"*.

(6) Treat "feet" or behavioral resistance as an issue of habit. *"Behavioral resistance has both systematic and cultural components"*.

(7) Diagnose the organization's resistance profile. *"Change what only needs to be changed"*.

(8) Activate the organization's own change cycle. *"Organizations are capable of changing themselves, once they are properly stimulated"*.

(9) Prevent resistance from ever arising. *"Help the organization learn the "Many-Few-Many-Few" process"*.

(10) Remember that organizational change is based on "pull through" not "push through" thinking and design.

DeLuca identifies three types of resistances viz. intellectual resistance, emotional resistance and behavioral resistance.

8.6   John P Kotter's "eight steps to successful change"

John Kotter's highly regarded books "Leading Change" (1995) and the follow-up "The Heart of Change" (2002) describes a model for understanding and managing change. Each stage acknowledges a key principle relating to people's response and approach to change, in which people see, feel, and then change: Kotter's eight step change model can be summarized as follows:

(1) Increase urgency – Inspire people to move, make objectives real, and relevant. The people should realize that the time has come for them to change and they cannot postpone it any further. Unless urgency is felt, work shall not get started at all in a number of cases, especially when we talk of implementing good management systems. People concentrate only on that day's production, rather than making a system better to get the results all the time without any hindrance.

(2) Build the guiding team – Get the right people in place with the right emotional commitment and the right mix of skills and levels. The section in-charges should be involved in the team as they have power for implementation. One should convince them and make them committed rather than making a team with juniors, just because they are enthusiastic.

(3) Get the vision right – Get the team to establish a simple vision and strategy; focus on emotional and creative aspects necessary to drive service and efficiency. Let the people realize their importance, and the target need to be achieved in order to make them successful in their assignments. If the vision is not correct and goals are given by referring to some other company without understanding the cultural changes required, the team might fail.

(4) Communicate for buy-in – Involve as many people as possible and communicate the essentials, simply, and to appeal and respond to people's needs. De-clutter communications – make technology work for you rather than against. When more people are involved, bringing change shall be easy and fast. Therefore educating as many people as possible regarding your theme, project, and system is essential.

(5) Empower actions – Remove obstacles, enable constructive feedback and lots of support from leaders (reward, and recognize progress and achievements). Allow the people to try their ideas to make the project a success. If people are free they get better ideas, and some times, much better ideas than what you can get as a leader or owner of a company. Much of the failures seen in textile mills are for not allowing people to try their ideas but to work only as per instructions of a boss.

(6) Create short-term wins – Set aims that are easy to achieve; in bite-size chunks and manageable numbers of initiatives. Finish current stages before starting new ones. Do not give a big target. Go on enhancing the target as and when you achieve.

(7) Don't let up – Foster and encourage determination and persistence, ongoing change, encourage ongoing progress reporting, highlight achieved and future milestones. Do not let up in case of non-

achievement. Analyze with the people involved and let them realize where they failed and what would have brought them the result. Firing a man for non-achievement shall make others to lose interest, as they all have worked hard as per the instructions.

(8) Make change stick – Reinforce the value of successful change via recruitment, promotion, and new change leaders. Weave change into culture. Do not allow to slip down.

8.7 SDI

Systematized Direct Induction (SDI) is a useful method for tackling "people issues". Workshops involving from 4 up to 100 individuals are organized, using members of same or different departments. This method addresses issues that members of staff may have with 'change'. Involving staff at the planning stage, allowing them to put forward their ideas, and preferred conditions, etc., makes the implementation of 'change' somewhat smoother. A planning meeting held by an elected staff member and a small group of the organizational staff will outline the problem/issue and decide if which staff needs to be attended. They need to ensure that the stakeholders are suitably presented at the meeting that has following steps:

(1) Initial introductions – To encourage inter-departmental mixing and supervisor/supervisee combinations, all staff members are encouraged to sit at the tables of four. The problem to be addressed is described and displayed, and participants are reassured that all suggestions will remain anonymous.

(2) Practice exercise – A specific colored slip of paper (say yellow) is issued to all participants. They are requested to 'identify their main issue in their daily work', write this on the colored paper, which are then collected.

(3) Identifying and discussing the key problems – Another set of differently colored paper is handed out (say red) with "how to" written across the top. Each individual must now complete the "how to" sentence with what they feel the company does that prohibits the workshop, sorting out their highlighted problem. Each table has a 5–10 minute 'buzz' session for discussing their thoughts.

(4) Identifying up to four more problems – Each participant completes another four red slips, completing the 'how to' sentence four more times.

(5) Ranking the five problems – Each participant now ranks their five problems, marking the most important pink slip '1' and the least important '5'.

(6) Break – Coffee/lunch taken, and during the break, yet more slips are placed on each table of a different color (say green).

(7) Cycle of generating and discussing solutions – After the break each participant selects their 'No.1' pink slip problem, and writes a solution for it on a green slip. Each table has a short 'buzz' session for discussing their solutions. This process is repeated for all five pink slips, creating five matching green slips.

(8) Workshop ends – Each participant clips their pink 'problem' slips and green 'solution' slips together in a cluster, and the workshop closes.

(9) Subsequent analysis – Each cluster is collected, collated, and analyzed to generate a management report. If the workshop was large, a small team may be required to do this. Incorporating company staff as well as external consultants will likely affect the final relevance and acceptability of any 'changes' that are implemented as a result.

8.8 Problem Centered Leadership (PCL)

The Problem Centered Leadership (PCL) technique identifies key requirements for someone facilitating a problem-solving group; its suitability is dependent upon the leader's sensitivity to group process. These behaviors cannot be applied mechanically. This technique was developed by Miner (1979) from the original ideas of Maier (1963). The technique outlines a particular scheme of stages, although it could be adapted to fit other stage schemes. It is summarized briefly below:

Stage	Suggested leader behavior
1. Presentation of the problem and relevant information	**Problem-Centered Leadership (PCL)** • Situation – not people related • Avoid suggesting solutions • Incorporate mutual interests • Include only one specific objective **Keep it brief:** • Present only essential clarifying information • Separate facts from interpretation • 5 minutes at a maximum

2. Initial discussion of the situation	**Give assurance to group members:** • Be realistic • Tell members that they do not have to accept any change **Allow expressions of feeling to be released in harmless channels:** • Look for guarded expressions of resentment • Leave long pauses to encourage expressions of feelings • Accept expressions of feelings • Understand, but don't evaluate, thoughts and feelings • Involve all group members in discussion
3. Continued discussion	**Minimal leader participation:** • Perhaps provide occasional summaries • Perhaps ask questions that raise still-unexplored issues
4. Solution generation and decision making	**Stimulate the generation of solutions:** • Prevent premature closure • Separate ideas generation from evaluation • Deal with agreement and disagreement • Summarize discussion periodically • **Assist in evaluating and selecting solutions:** • Examine pros and cons of each suggestion • Explore supporting evidence • Use stalemates constructively • Explore solutions for knock-on problems • Create short-list by voting and by combining choices • **Deal with disagreement by methods such as:**

	• Combining disputed options
	• Analyzing and trying to improve each separately
	• Treating failure to agree as a separate problem
5. Determination of decision acceptance	**Final leader summary:** • Provide a careful and detailed summary of the final decision • Ask group to check summary and modify as required

John Adair's Action-Centred Leadership model is represented by Adair's "three circles" diagram, which illustrates Adair's three core management responsibilities viz. achieving the task, managing the team or group and managing individuals. Good managers and leaders should have full command of the three main areas of the Action Centered Leadership model, and should be able to use each of the elements according to the situation. Being able to do all of these things, and keep the right balance, gets results, builds morale, improves quality, develops teams and productivity, and is the mark of a successful manager and leader.

John Adair observed what effective leaders did to gain the support and commitment of the followers. His model is important for two reasons: it's simple, so is easy to understand and apply, and he was one of the first to look at effective leadership from the point of view of those being led.

John Adair found that effective leaders pay attention to three areas of need for members of the team: those relating to the task, to the team itself, and to individual members of the team. At any time, the emphasis on each circle may vary, but all are interdependent and so the leader must watch all three. These are as follows:

(1) Task needs include setting a clear goal and objectives, and organization and management of the process.

(2) Team needs are the things like effective interaction, support, shared work and communication within the team and with other teams.

(3) Individual needs will of course vary from person to person, but the effective leader will pay attention to, and deal with, how each person is behaving and feeling.

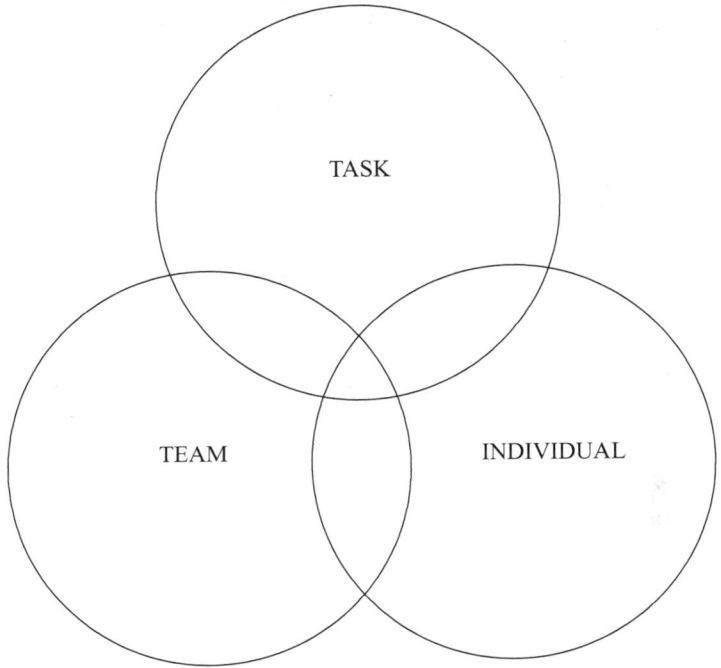

Figure 8.2 Adair's three circles diagram.

The three circles model is nowadays seen as rather basic, especially by managers who want to be considered sophisticated and up-to-date. However it's a good approach to learn early in your leadership career, providing a solid foundation for more complex human relations.

8.9   Responsibility for managing a change

Allan Chapman says that the employee does not have a responsibility to manage change; the employee's responsibility is just to do his best, which is different for every person and depends on a wide variety of factors (health, maturity, stability, experience, personality, motivation, etc.). Responsibility for managing change is with management and executives of the organization; they must manage the change in a way that employees can cope with it. The manager has a responsibility to facilitate and enable change; and all that is implied within that statement, especially to understand the situation from an objective standpoint (to 'step back', and be non-judgmental), and then to help people understand reasons, aims, and ways of responding positively according to employees' own situations and capabilities. Increasingly, the manager's role is to interpret, communicate, and enable; not to instruct and impose, to which nobody really responds well.

Allan Chapman further says that expressions like "mindset change", and "changing people's mindsets" or "changing attitudes", is not a good language for a manager, because this language often indicates a tendency towards imposed or enforced change (theory x), and it implies strongly that the organization believes that its people currently have the 'wrong' mindset, which is never, ever, the case. If people are not approaching their tasks or the organization effectively, then the organization has the wrong mindset, not the people. Change such as new structures, policies, targets, acquisitions, disposals, re-locations, etc., all create new systems and environments which need to be explained to people as early as possible, so that people's involvement in validating and refining the changes themselves can be obtained. He observes that whenever an organization imposes new things on people, there are difficulties.

Workshops are very useful processes to develop collective understanding, approaches, policies, methods, systems, ideas, etc. Staff surveys are a helpful way to repair damage and mistrust among staff. Management training, empathy, and facilitative capability are priority areas. Managers are crucial to the change process. They must enable and facilitate it. Merely conveying and implementing policy does not work.

Manager should not impose change; people and teams need to be empowered to find their own solutions and responses, with facilitation and support from managers, and tolerance and compassion from the leaders and executives. Management and leadership style and behavior is more important than clever process and policy. Employees need to be able to trust the organization. He insists that the leader must agree and work with these ideas or change is likely to be very painful, and the best people will be lost in the process.

We need to bring change not only for the benefit of self, but also for all those involved. Then the problems shall be eliminated on a permanent basis. We need to take responsibility for this.

8.10   Change management principles

Following principles given by Businessball.com should be always kept in mind while attempting and managing a change:

(1) At all times, involve and agree support from people within system (System = environment + process + culture + relationships + behavior, etc., whether personal or organizational).

(2) Understand where you or the organization is at present moment.

(3) Understand where you want to be; when, why, and what the measures will be for you to reach there.

(4) Plan development towards achieving the goal in appropriate measurable stages.

(5) Communicate, involve, enable, and facilitate involvement from people, as early and openly, and as fully as possible.

(6) Strong resistance to change is often rooted in deeply conditioned or historically reinforced feelings. Patience and tolerance are required to help people in these situations to see things differently.

(7) The psychological contract is a significant aspect of change, and offers helpful models and diagrams in understanding and managing change potentially at a very fundamental level.

(8) Certain types of people who are reliable, dependable, steady, habitual, and process oriented types often find change very unsettling. People who welcome change are not generally the best at being able to work reliably, dependably, and follow process. The reliability and dependability capabilities are directly opposite to character traits to mobility and adaptability capabilities.

(9) The more you understand people's needs, the better you will be able to manage change.

(10) Be mindful of people's strengths and weaknesses. Not everyone welcomes change. Take the time to understand the people you are dealing with and how and why they feel like they do, before you take action.

Bringing a change is not a simple task. To bring change, the leader has to change himself first. If the leader is not ready to change, no change shall take place in the organization. In a number of textile mills, it is seen that the leaders are talking of change, giving lectures about change not only in their mills but also to the outsiders in conferences and seminars, but are not able to bring change as they themselves are not confident about the new system. The management makes a slogan of "Customer delight" but does not respect the complaint given by a genuine customer. They talk of team building, but are working for breaking the teams that have been formed as they are scared that a new union would come up. People talk of employee-friendly policies, but harass their workers who are sincere and hard working. This is the reason why suggestion boxes are not getting any suggestions; the people feel that management is not worth trusting. They are always on the look for another job; the moment they get a slightly higher pay or a feel of security, they jump off.

8.11   Human factors for problem solving

Managements are increasingly concerned with the issues of responding to changing conditions and achieving higher level of productivity. Scott G Isaksen insists that the management must become more efficient in using the human resources to convene them to deal with these situations.

Problem solving is innovative and calls for novel and useful responses. Human factors related to individual styles, preferences to certain activities, motivational variables, aspects of group dynamics and development are important to seek improvement in productivity and effectiveness of problem solving.

There are varieties of ways to apply innovative approaches for dealing with people for problem solving. The first is to discuss with individuals and use the methods, skills, and techniques individually. The next is to work in small groups, setting up organizational structures and contexts conducive to problem solving innovation. Structures can be actual work groups or ad hoc task teams convened for specific task. It can be a self-managing team or a team affiliated to a larger team.

Process oriented skills are generally preferred which are durable and long lasting compared to specific technical skills. Technical training may be provided when that skill becomes a part of problem-solving exercise.

Scott G Isaksen suggests minimum five areas to consider while utilizing human resources for problem solving. These are orientation and readiness issues, how various levels of application calls for different leadership, how problem-solving challenges require diverse role for group members, aspects of group selection and development, and meeting certain logistical needs to optimize the problem solving.

Creative thinking and critical thinking are two separate issues, but have to be together for an effective problem solving. Understanding your type of thinking and problem-solving strengths and weaknesses can help in selecting a balanced approach and technique. Establishing clear ownership for directions and approaches can increase involvement and commitment. Ownership refers to the level of influence someone has in finding a solution to the problem. It also means that the individual is serious in implementing the solution. Ownership suggests that individuals are really looking for imaginative thinking and not just manipulating the same which are predetermined bias.

There are at least three levels in developing and implementing a problem solving process. The first is learning basic methods and techniques. The second level is practicing the process. The third level deals with real life concerns and opportunities.

The person who has the greatest degree of ownership of the challenge or problem is called the client. The person who helps the client in using problem-solving process is called the facilitator. Those who provide input and ideas are called resource groups. Understanding clear ownership or clientship reinforces the separation of roles, which is essential for problem solving. The facilitator guides the client and resource group in applying problem-solving techniques. The client makes decision and provides background information. The client ultimately evaluates and determines how much to be implemented and what was achieved.

The style of leadership should be consistent with group participation. While selecting a group factors such as number of members, diversity of people, jobs, ages, levels of expertise; appropriate people to play the role of facilitator, client and resource group need consideration. Normally, five to seven people shall make a team including client and facilitator. If the client is searching for an unusually novel approach, bring together a diversity of people who do not have a great deal of expert knowledge regarding the challenge or those who do not see it in the same angle as that of the client.

One of the classic leadership dilemmas is to get the work done while maintaining human relations. Understanding groups increases efficiency and productivity. The main problem we see in textile and garment industries is the conflict between the boss and subordinates while doing a work, although both of them are interested in the well-being of the company. Personal relations involve – how people feel about each other, the commitments they make, and the problems they have. Task functions involve learning what the task is, mobilizing to accomplish it, and doing the work.

While in the journey of problem the team undergoes four stages viz. Orientation, Dissatisfaction, Resolution, and Production as shown in Fig. 8.3.

Process of Team Building

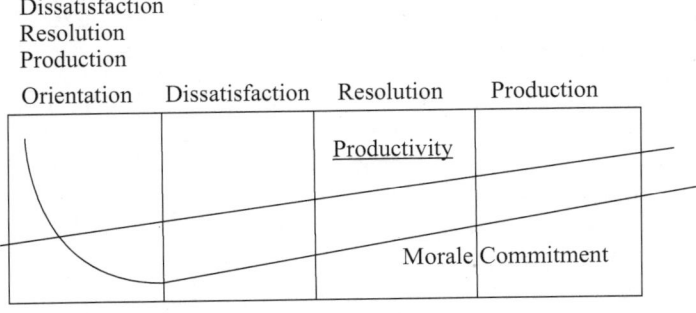

Figure 8.3 Stages of team building.

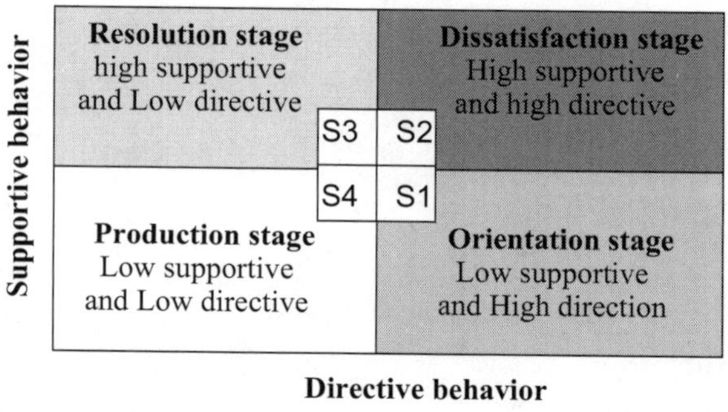

Figure 8.4 Change in leadership styles as the team progresses.

The leadership style also changes as the team moves from one stage to another as shown in Fig. 8.4.

The four stages of team building viz. Orientation, Dissatisfaction, Resolution, and Production are also referred as Forming, Storming, Norming, and Performing.

When the team forms, all will be enthusiastic and team morale shall be high. This is the stage of forming or orientation. When the work starts, people start facing hurdles and they blame each other for their failure; this stage is called as dissatisfaction or storming stage. A good leader observes the strengths and weaknesses of his team members and reallocates the jobs, which is known as norming stage or resolution stage. After this, the work becomes smooth and production starts. This stage is called performing or production stage. The leader has to change his style of leadership from stage to stage as explained in Fig. 8.4.

8.12   Problem solving strategies

The ability to be flexible is one of the most important aspects of problem solving. Everyone has problems which differ from each other. However, it is the strategies which a person uses to solve problems which will lead to either success or failure. There are a number of techniques one would want to employ while running into problems or challenges.

The first step to solve a problem is to look at any assumptions you may have about the situation. When many people are facing a problem, they make assumptions about the situation which can actually stop them from coming up with good solutions. Assumptions are essential as they set limits to the scope of the problem, and allow one to set up a foundation which can be used to solve them. Assumptions also display our desires. Assumptions can also make a problem much easier to solve. If a problem has no assumptions, it will be too broad to deal with.

An assumption should be based on laws of nature, not barriers that we create. When you run into a problem, you should take the time to think about any assumptions you are making about it. Is the assumption needed? Should it be used? The best way to answer these questions is to create a list of assumptions you have about a specific problem. Once you've done this, you should be able to look at the assumptions and decide whether or not they're based on facts or things you assume to be true.

Many people struggle to solve problems because they use assumptions that are not based on logic or facts. Just because you believe something to be true, doesn't mean it is. For example, the design of swimsuits for women were limited for many years because designers "assumed" that there were a limited number of materials which could be exposed to water, and they felt that most women who purchased swimsuits would swim in them. However, a designer did a study which showed that over 80% of women never wear their swimsuits in water and this opened the doors for using a wide variety of different materials for swimsuits.

Money is a major problem for many people, and they often make assumptions about it that are not based on facts. For example a person will assume that it will cost a certain amount to obtain something, without taking time to see if their assumptions are true. A good problem solver would ask, "How can I get the money to accomplish this task?" As you can see, they don't limit themselves.

If you are working with a team, cooperation is important. However, few people make false assumptions in this area as well. For example, you assume your team will be interested in using a certain strategy, or you assume that they won't be interested in a certain strategy. These are false assumptions which are not based on facts, because you have not got the opinions or thoughts of your teammates.

In a number of cases we assume that others are also looking at the problem in the same angle, what we are seeing. It is wrong. There may be someone who is not at all seeing this as a problem, but this as an opportunity to prove his worth.

When many people attempt to solve a problem, they make the mistake of placing emphasis on either the problem or solution. To properly solve problems, it is very important to make sure that you take the right approach. The approach you use for a problem is often connected to the solution. The first thing you should be familiar with is an entry point. As the name implies, the entry point is the part of the problem that you will first want to focus on.

Many people have linear minds, and approach a problem from the front end. However, the entry point doesn't have to be the front end of a problem. It can be approached from other angles. Most of the problems you encounter will have multiple entry points, and each one you choose will bring about a variety of different solutions. Few of these solutions may be better than others. Most people choose the first entry point they encounter because it is the most obvious. Many people believe each entry point chosen will give them the exact same solution. In reality, this is not accurate.

A hypothesis is an assumed explanation for a given amount of information. A rival hypothesis is one which goes against your hypothesis, and is another explanation for the same problem. The use of a hypothesis often deals with interpretations, and this could be difficult because the problem or solution may be based on bias. You should never limit yourself to a single hypothesis when you are attempting to solve a problem. You may use emotion rather than logic to support your hypothesis, and in a situation like this, important data may be discarded.

In few cases we assume someone as resourceful for attacking a problem; whereas there may be some other better person in the organization, whom we have never given an opportunity.

I think we have gone through sufficiently large number of various techniques for solving problems in textile and apparel industries. Let us now have a look at few cases in the next chapter.

Case studies

After studying different reasons for problem generation and techniques for analyzing the problems and taking corrective actions, let us have few case studies. The reader can think of different techniques for analyzing and solving the problem than what was actually done in the case, so that they can be prepared to face such situation if it occurs to them.

Number of cases said here may be happening today also in any of the mills around the world. There is no exception as majority of the cases are repetitive in nature.

Case No. 1 – Drop in C.S.P of 12s combed bleached yarn

A spinning mill which had yarn processing exported Ne 12s Combed bleached yarn to Canada. When the yarn was tested before dispatching, it had C.S.P of 2400. After 6 months, the mills received a complaint that the yarn C.S.P was only 1400–1500 and not 2400. The mill had used Shankar 6 cotton and there was no question of getting such a low C.S.P even by mistake. The grey yarn had a C.S.P of 2250 and the bleached yarn had 2400, which was normal. This incidence happened between 2001 and 2004.

The quality assurance manager was not ready to accept the figure quoted by the customer.

The mill had a practice of keeping four baby cones from each bleached yarn lot as representative sample. The C.E.O asked the quality assurance manager to get those samples tested again. When samples were tested, they were shocked to see the readings between 1200 and 1300; much below the readings quoted by the customer. The bleached yarn from the running lots were in the range of 2300–2400 CSP, made by same cotton with the same parameters. This problem of low strength was not observed by the local customers who were regularly purchasing this yarn for making sports socks.

As the mills could not find the root cause for this problem, it requested the customer to suggest some good laboratory where these yarn could be analyzed;

and finally, the Yarns were sent to a laboratory at Germany. The laboratory identified traces of iron oxide, which was reducing the yarn strength. High content of iron in the yarn was told to be the reason.

The mill is situated by the side of a river and they always get good fresh water from the river. The running water was tested and no traces of iron were found.

After making detailed analysis of the period of manufacture for that lot, it was found to be produced in the last week of May and there were no rains in that year. There was a big dam 25 km before the factory where the water level had gone very much down as compared to any year before and after the construction of the dam in 1965. The water received during that time was from the lower part of the reservoir, which was stagnant for a number of years. The fresh water used flowed only on the surface or dam, and the water lying at the bottom was always stagnant.

The local mills did not observe the drop in strength as the yarn was being used within 5–15 days, whereas in case of Canada, the time taken to reach itself was 5 months. The yarn was getting damaged gradually.

After this incident, regular checking of feed water for traces of iron was started, not only in that mill, but also in number of other mills, as this mill published their experience in various journals and also presented papers in conferences.

What was the root cause of this problem? Whether it was water getting contaminated with iron, or the technicians in the mill did not have the requisite knowledge of what all problems that could arise and how to be prepared to face the situation? When there are no rains, and the reservoir is becoming dry, it is natural that we get water from the lower level, which shall be more contaminated. Similarly, when there is excessive rainfall, lot of unwanted things may enter the river, which was otherwise buried in the soil long ago. It is needed to check the water in both the cases. A proactive approach of checking and verifying was not there and everything was taken for granted.

Case No. 2 – Steam damaging of grey yarn

A spinning mill was supplying 20s K warp yarns in both cones and hanks. A complaint was received by the mill from a customer located at Sholapur stating that the cones were steam damaged. The customer sent two cones as representative of the problem. The mill production manager and the quality assurance manager argued that the problem was not because of the mill, as there was no boiler or steam pipes of any type in that mill. When there are no steam pipes or boiler, how steam damage can take place? They argued that the housekeeping in Sholapur weaving factory was very poor, and the customer might have kept the cone bags exposing to steam either in his sizing area or at any other place. The management did not accept the complaint and wrote a letter to the customer that he was responsible for the loss. That customer stopped procuring yarn from this mill. This incidence was somewhere between 2000 and 2002.

After about 2 months, a similar complaint came from Panipat for the same yarn and the customer had sent four cones as representative of the problem. The management asked the quality assurance manager to go deep and find out the root cause of the problem. He went round the complete factory and godown, and gave a clean chit that there was no chance for the yarns to get damaged. However, the management was not ready to accept the theory because two customers from different places made the complaint, and there must be something happening in the mill itself.

After about 15 days there was a third complaint again from Panipat but from a different weaver, who was purchasing reeled yarns. The sample was very clear; half of the hank was damaged and remaining was very good indicating that the cones were damaged before reeling. Hence the question of yarn getting damaged at customer's end was ruled out.

The mill was getting the reeling work done on contract basis with an outside reeling factory. Both production manager and quality assurance managers went to reeling factory and showed them the hanks. The reeling factory manager told that they were getting number of cones like that, and it was not new. When he was asked as to why he did not complain to the mill, he said "You are the producer and you know your quality as you are checking the yarn and sending to us. We are only reeling whatever is sent by you. We thought that you are producing this type of yarn with some intention."

After hearing the statement of reeling factory manager, the C.E.O of the mill called all the staff and workers and showed the complaints received and requested all to observe and find the root cause.

After two days, one loading worker came to the quality assurance manager and asked him to go along with him as he had some clue. He took the quality assurance manager to the yarn godown and asked him to stand in the center and observe. The quality assurance manager could not observe or feel anything. The worker asked him to remove his shoes and socks and stand with bare feet. After removing the shoes and socks, the quality assurance manager could observe that the floor was hot at that place.

The godown floor was made with stone blocks joined by mortar. The mortar was damaged at a number of places.

The godown and the mill were on rocky land. At about 20 feet distance from the godown, a steam pipe was going at a height of 14 feet and there was a steam trap. A big banyan tree was there and lots of dry leaves had fallen and blocked the steam trap. As the area was covered by a compound, it was not visible that the steam trap had blocked. The condensed hot water was percolating down in the earth.

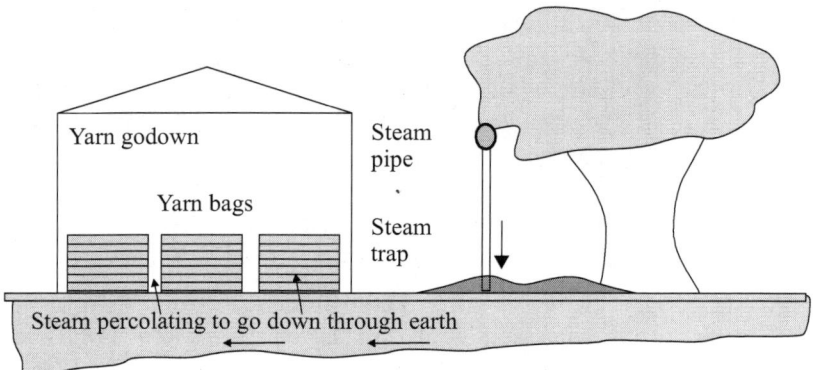

Rocks below the ground level

All the yarn bags in that godown were removed and opened. There were nearly 300 cones which were damaged due to the steam in the bags (those which were kept on floor).

Arrangements were made to repair the floor and the steam pipe was shifted to a different place. The worker was rewarded for his initiation in identifying the root cause of the problem.

What is the root of the problem? Whether it was the steam percolating from ground, or the negligence of the staff in not removing the leaves and cleaning the floor, or not maintaining the floor where the mortar was found damaged, or installing a steam trap without understanding the ground conditions? Further, the attitude of the technicians in rejecting a complaint just because they were not able to identify the root cause is very dangerous. One should understand that no customer is interested in making a complaint but is interested in running his business smoothly. Whenever he makes a complaint, we need to go deep and analyze the cause.

Case No. 3 – Polyester contamination in 100% cotton spinning mill

An export oriented spinning mill in Karnataka producing 100% cotton grey yarns received a complaint of polyester fiber contamination. It was a surprise as no polyester was brought to that mill at any time. Also, the mill in order to avoid the complaints of Jute contamination and HDPE contamination were insisting on covering the cotton bales with grey cotton cloth. Bales covered with Jute or HDPE were not accepted. How polyester fiber contamination can come in their product was a big question. This incidence was somewhere between 1998 and 2001.

The quality-control checkers were asked to check all material in the mill. A Q.C. investigator collected all the cloths available in the mill including bale covering, green clearer cloth used for roller covering, the sider bags used for collecting bonda wastes, the bags used for collecting the soft wastes, and analyzed them for the presence of polyester. They were shocked to see cotton bales covered with polyester cotton blend with 65% cotton and 35% polyester. As cotton was costlier than polyester, some intelligent fabric manufacturer had procured 35:65 PC cloth as it looked more like cotton yarn and the cost was low.

After this incident, the mills started checking all the cotton bale covers using hand held UV lamps to identify the possibilities of PC fabric or any other fiber coming. Letters were written to Textile Commissioner's office and articles were published in major textile magazines regarding the cheating being done by weavers.

The root cause was found to be the greediness of the weaver who sourced PC yarn just because it was cheap, and the ginning factory that was not bothered to check the fabric procured in spite of the EOU unit instructing the ginning factory that they wanted bales to be covered with pure cotton grey fabric.

Case No. 4 – Jewelry found in cotton balls

An exported oriented cotton non-woven plant in Maharashtra, engaged in the manufacture of surgical pads, cosmetic pads, cotton balls, ear buds, absorbent cotton, etc., received a complaint stating that a piece of jewelry was found in one of the packages of cotton balls made from bleached cotton. The complaint was from a famous company from UK which sent the photo of the ear ring made of gold. The party claimed £20,000 as compensation for this negligence as the materials were being used by surgeons for the purpose of surgery. The poor girl, who lost her golden ear ring, did not dare to tell anyone that she lost her ear ring while working, as it would have costed her job also along with the lost ring. This incidence was in the year 2005–2006.

After this incident, the company banned all the ornaments worn by anybody except for the '*Mangala Sutra*' which should be fully covered.

Case No. 5 – Quantum of contamination

The problem of contamination in the cotton yarns exported was a burning problem for Indian cotton industry for a long time, and a number of measures were taken to overcome it. One such measure was engaging ladies for sorting out contamination from the cotton bales. Normally, one bale per lady was the workload. The ladies were asked to show the contaminations picked by them, weigh it, and make a record against the bales number. This system was developed during 1995–1998 and was well-accepted as a good practice till the bale pluckers and online contamination clearers were developed and installed.

In a mill where a number of ladies were working, a supervisor was monitoring their activities and ensuring that everyone picked as much contaminations as possible, and was reporting to spinning master. One day in a particular cotton lot, the contaminations were very less compared to other lots. But the spinning master was not ready to accept the fact and started firing the supervisor and the lady who sorted the contamination from the cotton. He said "Unless you take out 300 gm of contamination per bale, I am not going to keep you on work".

There were continuous efforts at ginning factories and ATIRA was guiding ginning factories to make the cotton clean; and hence, in the lot produced from such ginning factory had less trash and absolutely no contamination. But the spinning master was not ready to realize the situation. The supervisor and the ladies got fed up with the spinning master and started bringing few old torn cloths from home. They also went to cotton waste yard and collected back the contaminations deposited and made 300 gm bags, put the bale numbers and showed the same to the spinning master. Whenever the contaminations were more, they were kept aside for emergency purpose by which the purpose of weighing and recording against the bale number and lot number became a mockery. Management never got the information as to which ginning factory was performing well.

An irrational target makes people to cook the figure, and the company cannot identify the real problem.

Case No. 6 – Over production in sewing in garment industry

In garment industries, target for sewing production is fixed after conducting time study. However, few operations get missed while conducting the time study and the garment factories always produce very low efficiency than what is fixed as standard. In one such case, when the production was very low as compared to the standards specified, the production manager insisted the supervisors to record more production for the day and adjust it from next day's production. However, the production in the next day was also less and the low production continued as there was no system of identifying the root cause of the low production. Suddenly the lot was completed as per the production record, but still a number of panels were in pipe line. In between, the supervisor was changed and the production clerk did not allow the next day's supervisor to record low production as it would reflect bad on his performance. In the end, huge over booking was observed. Then the production manager insisted the quality checker to make a fake inspection statement and reject the excess garments accounted as rejected and sent for rework. Records of rework were to be made to balance the efficiency. The real reason for low production was the wrong target fixed.

The costing done was wrong because of the wrong prediction of productivity, and unnecessary cooking of figures was done while masking the actual reasons for loss. If actual reasons were shown to management, they would have taken few corrective steps and avoided this type of problems.

The root cause is lack of knowledge of time study. In a number of garment industries, it is seen that people without having requisite educational background and training in industrial engineering for conducting time and motion study, and production forecasting by the time study are engaged as industrial engineers. Similarly the production managers engaged are just senior workers who do not have any formal education and training for production management, line balancing, activity mapping, etc. The garment industry is facing a problem that qualified people are not ready to work in production area which is monotonous. They want either merchandising, or designing, or fabric sourcing. The companies are also not spending on providing training to existing staff in order to enhance their knowledge and skills to perform their jobs better. The people are always under stress for giving deliveries.

Case No. 7 – The Tamba Kaata Experience

It is case of the year 1988. A small spinning mill from a rural place, Venus Textiles, which had no marketing set up was engaged in the production of polyester viscose yarns and selling it to Bhiwandi market through a trader at Tamba Kaata in Mumbai. The trader named Mahendrabhai was taking the yarns from small spinning units and stocking at his godown at Bhiwandi and selling to small weavers in retail. The weavers were taking two to ten bags at a time by paying cash. Mahendrabhai was taking yarn from the mills on credit basis and was paying them after 2 months.

The production incharge of Venus Textile Muthuswamy had come to Mumbai for some work. The owner of Venus told him to meet Mahendrabhai and remind him of the payment due to the mills. Mr. Muthuswamy took the address of Mahendrabhai and was able to locate his office, in a small room on the third floor of an old building, a table shared by three people, and a common telephone. Mr. Muthuswamy introduced himself to Mahendrabhai, who greeted him. He called a boy and told him loudly that "Manager of Venus Textile has come, bring two special teas". The boy went down to bring tea and came with tea after 30 minutes.

When the boy had gone out for bringing tea, Mr. Muthuswamy was asked to sit and Mr. Mahendrabhai was looking at few urgent papers. Then a phone came, Mahendrabhai took it and answered. It was like this:

Mahendrabhai – "Hello. Mahendrabhi is talking".

Other side "-------------------"

 Mahendrabhai – "What? 34s yarn is not working well!"

Other side "-------------------"

Mahendrabhai – "Which lot? Oh! Of Venus!"

Other side "-------------------"

Mahendrabhai – "What is the problem?"

Other side "-------------------"

Mahendrabhai – "Simply not working?"

Other side "-------------------"

Mahendrabhai – "You want to return the material! How can I take it! You have opened the materials and consumed half."

Other side "-------------------"

Mahendrabhai – "This time you manage."

Other side "-------------------"

Mahendrabhai – "Loss! What loss!"

Other side "-------------------"

Mahendrabhai – "Okay I will talk with mill owner! Please use it!"

Other side "-------------------"

Mahendrabhai – "I am guaranteeing! Arey don't worry."

Other side "-------------------"

Mahendrabhai – "I shall get you discount. I shall talk to owner."

Then Mahendrabhai turned to Muthuswamy and said "Sir, it has become a headache to really sell your yarn. My time is spent only in answering these people.

Within 6 minutes there was another call by another customer complaining about the performance of 40s PV of Venus. Same dialogues repeated as that was for Ne 34s.

Mahendrabhai just reviewed 3 papers and one more phone call came. Another person sharing the table took the phone and gave to Mahendrabhai telling that "All calls are for you only. Shamser Singh is complaining on 20s PV of Venus". Again the discussion went on.

In the one hour time during which Mr. Muthusawmy was sitting in the Mahendrabhai's office, there were five such calls, two complaints on 34s, two on 40s, and one on 20s, and all were of Venus.

Mahendrabhai said "Sir, believe it or not, I am not able to take food or even sleep as these people shall be continuously calling and asking me to take back the yarn. You only tell me what I have to do. I know I have not paid your company for two months, but how can I pay. I can pay only if your yarn is sold."

Poor Muthuswamy returned without saying anything. When Muthuswamy narrated the incident to his owner, he said "Dear Muthu, being a technician you are only fit to get production. These traders create drama. How can there by so many calls, all about our yarn when he is dealing with yarns of a number of companies? Why all the calls came when you were sitting in his office? If selling our yarn was so difficult, why he is lifting all the materials by sending his trucks in time? The phone calls were made by that boy who went with the pretext of bringing tea, and called from public booth just below his office. He cannot make this drama in front of me. I shall go next week."

What we see and what we hear might not be a fact but an illusion. One should have maturity to differentiate them.

Case No. 8 – Hipparkar

It is a case of the year 1979–1980. Suddenly a police van came to Madhavnagar Cotton Mills in Southern Maharashtra at 4.30 PM, the starting time of the second shift, and enquired for one Mr. Hipparkar, a weaver. The time office person informed that Hipparkar had gone in for work. Two police officers went inside, whereas other three were watching the gate.

When the police entered weaving shed and met the supervisor Mr. Deshpande, he told that just 2 minutes back he took leave stating that he had

Solutions to problems in the textile and garment industry

stomach pain, and must be going out through the gate. But Hipparkar was not traceable. The police arrested Mr. Deshpande alleging that he was a part of Hipparkar's team who were engaged in robbery in Tamilnadu–Andhra boarder between Chennai and Nellore.

Hipparkar was a weaver who was very rich as compared to other weavers. He was having a Bullet motorbike and own house. He was regularly taking leaves in the first week of every month. The police asked Mr. Deshpande as to how he sanctioned leave for Hipparkar when he was not entitled for a leave. Deshpande explained the practice that whenever there were sufficient workers to work, leave was given to the people who asked. As there were 30% substitute workers, there was no scarcity of weavers. The management was paying leave wages only as per their eligibility and other days were treated as leave without pay.

The police took an objection and said that "You are supposed to give leave only as per their entitlement and should refuse when they do not have leave in their credit. The police also asked him as to whether he had enquired Hipparkar as to why he wanted leave and why he was taking leave in the first week only.

As per the standing orders, the leave without pay was to be sanctioned by the Mill Manager, whereas it was not done. Mr. Deshpande told that he was not aware of the rules, for which the police objected while asking him how he could become a supervisor without knowing the rules and regulations. The prime responsibility of a supervisor is to follow the rules and regulations of the country and the state.

Mr. Hipparkar and his team were involved in the robbery in Chennai–Nellore highway near the border. One of the victims identified Hipparkar in a goldsmith's shop and tipped the police. As the police went late to the shop, they could not get him there but got his address as he was regularly visiting that goldsmith. The Andhra Police came to Maharashtra and went to his house first and found that he had just left for work.

The police asked Mr. Deshpande as to who gave message to Mr. Hipparkar that police were at gate enquiring for him. Mr. Deshpande had no answer as he was not aware of anything. The police alleged that there was a good communication system of sending signals about the movement of police and hence, Hipparkar get the message and escaped by jumping out of the compound wall and did not go out from the main gate. He had no stomach pain. The police went to his house again just to see the house locked. Hipparkar and his wife both had escaped on their bike, and they were never traced by police.

The management did not want to come in the picture and simply dismissed Mr. Deshpande.

An important lesson for technical staff is that "Do not cross the limits. Understand the rules and regulations and never violate them".

Case No. 9 – Taking order of 10CHC in mill running on 100 C Weft

This was somewhere in 1978–1980. The mill was regularly spinning fine counts with an average count of Ne 60s. They had 50,000 spindles, out of which 32 ring frames of 408 spindles were of 5"lift, 1.25" ring dia and with 3 leaved weft cam, suitable for producing direct weft cops. The mills were supplying direct weft cops in Ne 80s and Ne 100s for decentralized power looms, manufacturing dhotis in southern Maharashtra. They had 20 Naismith Combers producing around 125 kg of combed material per machine per day.

Once a customer dealing with yarns for embroidery wanted 10,000 kg of Ne 10s Combed and was searching for a good spinner. The customer gave an offer of Rs 5/- profit per kg of yarn, which was very lucrative.

The Managing Director asked the Project and Development officer, whether he could produce the yarn of the required quality. There were 9 ring frames with 8" lift and 1.75" ring diameter and they were suitable for coarse counts. The officer with the help of spinning master produced a sample and submitted, which was accepted by the customer. The parameters of speed, expected production per shift considering the number of doffs and loss of efficiency was calculated and costing was done. The customer happily agreed to the cost and asked the mills to add Rs 5/- per kg as their profit and placed the order.

As the order was very lucrative, the management wanted to take production at the earliest. One ring frame was set for this new production, and the materials from comber were diverted to this machine. The working was good. The problem started after 24 hours, when the worker from 100s Ne machines came and told that all those machines were running short of material as all comber materials were being consumed by this one machine. The production per spindle in Ne 100s was 18 gm and in 10 CHC, it was 360 gm. One spindle on 10 CHC was equivalent to 20 spindles on Ne 100s in production in grams per spindle.

The weft ring frames were not suitable for any count coarser than 80s, and only combed materials could run there as there was no demand for carded yarns in 80s and 100s. Hence, till the order of Ne 10 CHC was completed, 20 ring frames of Ne 100s were to be stopped, which was never considered while making costing for 10 CHC. The dream of a profit ended up in loss.

Root cause of the problem was not reviewing the expected problems because of introduction of a new product, which is now strongly recommended in the clause of ISO 9001:2008 in 7.3.4.

Case No. 10 – Holes in the fabric seen after processing

This was somewhere in 1979–1981. In Madhavnagar Cotton Mills, a composite mill in southern Maharashtra, the folding master rejected lots of fabrics for holes in almost all sorts which were processed, finished, and ready for packing. The dyeing and finishing masters blamed it on grey fabrics and said there was no such process which could make holes in the fabric. Also, how can there be holes in all the varieties when they have undergone different processes was the question. The weaving master challenged and asked the processing people to check the grey fabric before taking it for processing, if they had doubt on weaving. He also told that the holes could not be in all varieties as they work on different looms and there were two sheds, one for auto looms and other for plain looms.

The weaving master personally checked all the 340 looms and convinced that no holes were being generated at loom stage. As Project and Development Officer, I also took rounds and went to each and every corner, wherever fabrics moved. I could not find any clue for the problem.

The S.Q.C persons were told to check all the processes but could not come with any possible reason for this problem. There were no holes in any of the grey fabrics, whereas it was seen only in the finished materials.

After listening to the problem, the Spinning Master, (Late) G. V. Anturkar came forward and asked permission for probing into the problem as he was a neutral person and in no way had link to the problem. After going through all the places where the materials were kept, moved, or processed, he decided to come in the night and study again.

Mr. Anturkar came at 11.30 PM, half an hour before the shift change, and was shocked to see burning particles coming out of chimney from the boiler. The fabrics to be sent for processing were loaded in hand trolleys and were waiting in the door of process house. Some burning particles were falling down, but were getting extinguished before they reached floor, and no one could feel it.

Mr. Anturkar immediately alerted the process-house staff and moved the trolleys away as there were chances of burning particles falling on the fabric. The arrangements were made to cover the trollies with tarpaulins.

This problem was not there earlier, as the mill was using only coal for firing boiler. Recently they had switched over to bagasse as there was shortage of coal. The bagasse being very light was being carried away by the flue gasses. The burning particles could not be seen in day time, and also in the night it was not continuous.

Case No. 11 – Exhibition of barre effect

This was in the year 2000, when almost all spinning mills supplying yarns to hosiery were facing the problem of barre effect in knit goods. Gokak Mills was also not an exemption. Lots of studies were made to find the reasons for barre and over 140 reasons were listed. The mills had a practice of preserving all the complaint samples with labels so that the same can be shown to the staff and workers repeatedly.

Gokak Mills conducted a seminar on barrie with the theme "Barre the Barrier" and as chief of Research and Quality Assurance, I was the main person organizing the seminar. Mr. H. S. Bhaskar, the Executive Director suggested display of all the complaints and the analysis made so that all can understand the reasons.

We had kept samples with barre effect for more than five years. When we opened the cartons and removed the samples, I was shocked to see that in more than 50% of the cases, the barre effect was not there, which was there earlier when the complaint was received and we had made analysis and had attributed various reasons like tension variation, loop length variation, probable mix of carded and comber yarns, variation in hairiness, probable chance of mix up of two lots of yarns, and so on. The barre effect had vanished as the cloths relaxed for a long time.

Case No. 12 – Cone sticker for 20s KHC (East European experience)

This case was somewhere in 2000–2002. I do not remember the customer's name that was from an East European country. The customer was a trader, and had booked order for 20,000 kg of 20 CHC yarn and had sent 20,000 labels to be put inside each cone. The label was not in English and none of our people were able to read it. There was a good sketch of cotton ball.

As the cone weight was 1.8 kg, we were left out with number of labels.

After 2 months, we received order for 8600 kg of 20 KHC from the same party. This time they did not send any labels. The production manager enquired with the marketing people about the label, and the marketing chief replied by saying that as already customer has sent enough labels, they have not sent any labels this time and asked us to go ahead with production and stick the labels in stock.

After 3 months of supplying the material, a complaint was received regarding wrong labeling. The customer claimed a big compensation for cheating. After detailed discussions we came to know that the label was for combed cotton counts and not for carded yarns. As the label read combed, the customers made a hue and cry as they purchased the yarn as combed yarn, and could realize only after the material was used and they did not get the required feel.

The problem was due to assuming and not enquiring with the customer even when a doubt was expressed by the production personnel. Why customer sent extra labels should have been enquired in the first order itself.

Case No. 13 – Visit of Mr. Barcroft

This was in the year 1976. Mr. Barcroft, an industrialist from Manchester, was visiting India in search of yarns for his factory. He had doubling and twisting factory.

In our mill we had 10 roving machines and 2 slubbers, which were idle after the inter frames were converted to WST UTM-620 high drafting. The management had taken a decision to scrap the roving machines as there were no takers for them as a machine, but the scrap dealers were keener because of the good quality cast iron in the machine. We had dismantled 4 roving frames and transferred them to scrap yard, but still were not sold. This was the time we got the message about the visit of Mr. Barcroft.

We were doing few trials for using these machines by producing waste yarns of count Ne 3s and Ne 4s. It was decided to close the roving shed and close the doors as we did not want our guest to get a bad impression.

We had produced a number of samples from Ne 20s to Ne 120s, and doubled yarns from Ne 2/20s to Ne 2/120s and had displayed them in a separate room that was normally used for seminars and conferences, and a small library was also located in it.

Mr. Barcroft was taken round the mill after showing all the samples. He was not impressed with any of the yarn samples showed to him.

When he came near the roving shed, which was locked, he asked me as to what was there in that room. I just told him that some old machines were there. He wanted to see them, and there was no other way for me therefore, I got the key and opened the room. Our director who was also with him during the visit was little bit upset as I took Mr. Barcroft in to the roving shed.

When Mr. Barcroft saw the waste yarn on roving, he was so happy and said "This is what I wanted and searching all over". He asked me, could I produce

Ne 4/3s. Our director Sharad babu told that we had no heavy doublers to deal with that type of yarn. Mr. Barcroft laughed and said "You have beautiful slubbers, reverse it and convert it as doublers". He gave the guidance to make a warp beam of 104 ends and creel it in the back of slubber, and take four ends each per spindle. By his guidance we were able to produce 4/3s of fixed length yarn using the roving and slubber machine, which was treated as great wonder. Mr. Barcroft explained that 100 years back that was the technology used and ring frames were not there. He showed old books and photos.

The management asked me to bring back all the dismantled machines and re-erect them. However, this product could not be run for a long time as Volkman VTS-06 started entering the mills and could give 28–32 times more production as compared to slubbers. The slubbers were running at 500 RPM and giving one twist per revolution whereas VTS 06 could run at 7000–8000 RPM and give 2 twists per revolution.

It was really a challenge to convert slubber into a twister running in reverse direction as we had to convert a number of mechanisms to run in reverse direction. It gave a lot of confidence to me and to the management to see that we could do something creative.

Case No. 14 – Collapse of a spinning mill supporting the family firms

A leading spinning mill in Southern Maharashtra was founded in 1944 and run by the four brothers of a family and was doing well for over 30 years. As the brothers were getting old, they decided to convert the mill from partnership concern to a public limited company and shares were issued to the younger family members like daughters, sons, son-in-laws, daughter-in-laws, etc. As the company grew, small units were started in the name of individuals who were either supplying or purchasing from this company. The managing director was always selective in dealing and was ensuring that the ancillary firms do not make loss. He was selling the comber noils to the surgical cotton plant at a much lower rate than that was quoted in the market; similarly, the weaving factories were getting yarn at a lower price. He was purchasing the materials at a higher price.

It worked like this for nearly 10 years, and every year the mother mill was showing losses whereas all the ancillary units were making profits. The youngsters in the family started thinking that the mill had become old and hence was not able to make profit, and it was unviable to run it. They started demanding the selling of the mill and distribution of the money among all family members as per their shares so that they can become independent and improve their units and grow. No one was ready to listen to the old managing director, who was the only person surviving among the four founders, as the children and grandchildren of other brothers had grown up. Finally, the mill was sold out in 1986. The managing director decided to come out of business and left the town and settled at a peaceful village and spent his remaining time in social service to that village people.

The new owners of the mill stopped all concessions given to ancillary units as they were no more part of the family of the mill. When the ancillary units had to procure materials at market rate and provide materials at competitive price, they realized the strategy followed by the old managing director. The ancillary units that were boasting of making huge profits while the mother mill was making losses started facing loss. As the owners were not exposed to competitive environment earlier, they could not face the competition and were forced to close down.

The policy of reservations and subsidies is more harmful to the society, rather than allowing people to become strong by exercising in the competitive atmosphere. Initially some help may be given, but as a loan and not as free gift.

Case No. 15 – Collapse of a spinning mill because of greediness of brothers working as directors

It was a case of 1971. Two brothers were partners and running a small spinning mill of 12,000 spindle capacity in Davangere District of Karnataka which was founded by their father who had died recently. The factory manager was a well-qualified technician and was well known for his sincerity and selfless working.

Among the two brothers, elder was looking after marketing and the younger purchases. There was a competition between the wives of the brothers to gather more riches for themselves. They were insisting their husbands to earn more.

As the brothers were not exposed to any scientific management systems and were not able to think and bring the mills up, they started using unfair means to earn for themselves by cheating the other partner. The elder brother was taking the yarn out without making any gate pass or entries in the registers, whereas the younger was demanding the factory manager to sign the receipt of cotton bales which had not arrived. The factory manager got fed up with the approaches of the two brothers and left the company. There is no need to say that the company collapsed shortly and is no more.

The greediness killed not only their business but also the future of the people who were depending on them.

Case No. 16 – Cuts in weft pirns

It was a case in 1975. There were only shuttle looms in India. Good mills had auto looms and for weft preparation automatic pirn winding machines were being used. The mill had a peculiar problem of cuts in the pirn produced on latest Schweitzer MSL pirn winding machines. The cuts were seen clearly on the pirn surface. How the yarn could run on a high speed pirn winding machine with such cuts was a question.

When detailed analysis was done, it was observed that the pirns wound in 'C' shift only was having this problem and it was not there in 'A' and 'B' shifts.

A careful study of work practices was done and finally the root cause was found out.

In third shift, one weft-sorting worker had got recently engaged and was wearing new rings in both the hands. One of the rings had sharp edges and when he was sorting and aligning the cops, the ends touching that portion was getting cut.

The mill banned the workers from wearing rings during the working hours.

Case No. 17 – 2/28K – SS and ZZ No complaint from a big customer but too many complaints from a small weaver.

A spinning mill was producing high twist yarns for crepe bandages. The counts were Ne 2/28s in SS and ZZ twist. A reputed medical appliances manufacturer of international fame was regularly purchasing 20,000 kg in each twist every month.

As manufacturing exactly 20,000 kg was not practicable, there used to be some 50–200 kg surpluses every month, which was procured by a small weaver in decentralized sector.

The multinational company purchasing the yarn in bulk never made any complaint regarding the quality of yarn; on the contrary he had given certificate of reliable vendor to this mill. The small weaver was sending complaints about the quality and was sending the cones as a representative of the problems.

The quality assurance manager was not able to understand this. The complaints given by the small weaver were genuine. One day he met the small weaver and asked him the possible reason why so far the big buyer had not made any complaint as the materials sent to this small weaver and him were same. The small weaver said that the products of that multi-national company were sold on the brand name itself and no one was questioning their quality or rate. For that company, the 20,000 kg of cotton yarn was nothing, as there other products were earning more. They can afford to keep the bad cones aside and destroy them, whereas being a small weaver he cannot do that. The customers of small weaver were very critical and checking each and every meter of the fabric. The small weaver cannot sell unless his quality is better than that multinational company.

This is a great lesson. One cannot assume that the products are good just because the customer is buying without making any complaint. One cannot boast of quality just because he is supplying to established brands.

Case No. 18 – Accident analysis

In a spinning mill in Southern Maharashtra, the management was taking a number of steps to reduce the accidents. They had a system of reviewing the number of accidents on monthly basis and analyze the reasons.

Once management decided to go more in detail and an analysis was made of the persons getting injured. The management observed that only certain people were meeting with accidents and the others had never met any accident. A detailed analysis indicated that the number of accidents were more on 8th and 25th day of every month. It was found fishy. The people normally meeting with accidents were summoned and told that they were not careful while working and hence were repeatedly meeting with accidents, and action shall be initiated if they meet with another accident again. Then one of the workers requested the management to send the wages by cheque to his house instead of handing over cash. There was drastic reduction in the accidents.

8th was the salary day and 25th was the day on which advances were given to workers. The workers were normally having some loans taken from private money lenders, who used to sit in front of the mill gate for recovering the interest and the principal amount. The action of crediting the amount to the account of the employees reduced the accidents and also the absenteeism after salary day.

Case No. 19 – Government policy of hank yarn obligation

The hank yarn obligation is a mechanism to ensure adequate availability of hank yarn to handloom weavers at reasonable prices. The existing hank yarn packing notification dated 17.04.2003 promulgated under Essential Commodities Act, 1955, prescribes that every producer of yarns who packs yarn for civil consumption, shall pack at least 40% of yarn packed for civil consumption in hank form on quarterly basis and not less than 80% of the hank yarn packed shall be of counts 80s and below.

The intention is clear. The government wants to help handloom weavers, especially those are in rural areas. Mahatma Gandhi propagated for development of *Khadi*, as India was under the clutches of British and the fabrics were being imported from Manchester. The cotton grown in India was being exported to Britain and the fabrics were being imported. There was unemployment in India and *khadi* was a means for providing employment. During 1955, when this act was made, there was a need to promote handloom to prevent import of fabrics. But now the industry has grown. We are not importing any fabrics but are exporting to world over. The handlooms are getting disappeared as weavers are installing high speed shuttle-less looms even in decentralized sector. Further, the numbers of spinning mills have increased and the yarns are also being exported. India is now the second largest producer of yarn. If 40% of its yarn sold in domestic market is made into hank, the industry cannot survive. It means the rules are on paper and documents are prepared to show that 40% are being sold as hanks. The officials are getting money from the mills for making false documents. The government is seeing only the documents submitted to them and have blind eyes on the statistics of number of handlooms actually working, the yarn consumption for them, the yarn produced in India, the yarn consumed by local market, and so on.

Case No. 20 – Modernizing led to loss

Numbers of spinning mills were modernized in the 1990–2000 era by installing latest high speed machinery in South India. They sold out the old machines. The number of people working in organized mills was reduced because of the automations. The workers were paid lucrative amount on VRS scheme. The workers purchased old machines, made their own teams and started small spinning units. Over 800 such small units came in Coimbatore District with 1000–5000 spindles. The traders gave them cotton, and units worked on job work basis. This added to the problem as there was already 35–40% over capacity in spinning compared to yarn requirement. The mills had to face the competition from these small units because of their lower cost of manufacturing and ability to take small orders of specialty yarns. The root cause of the problem faced by mills was their greediness. Instead of scrapping the old machines, they sold it to get higher returns and invited a permanent problem.

Case No. 21 – Over capacity of spinning

Till 1989, the Government of India was controlling the number of spindles installed in the country. When the liberalization policy was adopted and mills were allowed to install any number of spindles, the spinning capacity increased. Everyone thought that if he increased the number of spindles and total production, his cost of manufacturing shall come down, but never realized that the total yarn consumption depends on the requirement of customers. Although statistics were given by textile commissioner's office indicating that spinning capacity was much higher, people still invested on spinning while telling that they shall export the yarns. The total spinning capacity world over was more than 40% of total yarn requirement in the world. Even then the increase in spindles went on stating that "We will be fit with latest technology, and only old units shall close down". However, old and small units, which had very low overheads, low or nil interest, and depreciation burdens, and had name in the market from a long time, could survive the competition; but the companies with poor marketing abilities failed in the race.

Case No. 22 – Shortage of skilled people

The mills say that they have shortage of skilled people and hence production is suffering. Further, they also claim that market is very bad and there is no future for textiles. But none of them are ready to close their business and start some other, where they can get profit. They stick to textiles and whenever an opportunity comes, they try to expand.

Numbers of mills especially in South Gujarat area are paying very low wages compared to the minimum wages fixed, and also overtime wages are not paid as per Factory Act. If the minimum wages fixed by government is Rs 201 for 8 hours working, the workers are paid Rs 170–180 for 12 hour working. All the workers are migrated from UP, Bihar and Orissa. They come in a group, join some mill, work for 8–10 months, and go back. No efforts are made to provide good facilities for the workers to live there, bring their family, settle down, and train the people to suit the industry needs. The mill management feels that investing in training is a waste as the trained workers demand higher wages and jump to other mills. The management blames the workers for absenteeism and the government for the shortage of skilled workers.

There was a time when the unemployment was very high in India, but now the time has changed. Employment opportunities are high and lucrative salaries are given by other sectors like hotels, tourism, hospitals, IT, etc. Numbers of people are getting lucrative jobs abroad, and most of the people prefer going out to earn. Only few illiterate or less educated people, who have no confidence of going out are trying to stick to local industries. Even if the salary paid is high, numbers of people do not prefer to work in textile considering the working conditions.

Case No. 23 – Assuming others as fools

This is not a single case, but it is normally seen in almost all mills in India. Whenever a new man joins, especially in a higher post, people normally do not give him the correct information and try to mislead him. The person who joined also normally tries to impress people by big words and tries to prove what the earlier man did was not correct. In this process, 6 months shall be lost by the time people join hands and real improvement process starts.

People talk about team building, mutual trust, cooperation, etc., but when it comes for them to act, they act exactly in the opposite way and people doubt their moves. Majority of the problems in textile mills are due to mistrust among the people. The management thinks that employees are not trust worthy and therefore, no power is given to them. They try to control everything. The employees feel that top management is making money by them while ignoring the hard work of employees. There is a saying that "The accounts auditor tries to prove that except him all others are thieves". Similarly, a manager tries to prove that he is the only efficient person and others down below him are useless. Always blaming other person for the failures and claiming a success as because of him is a normal tendency; although, people by heart know what they are doing is not correct.

In a number of cases it is seen that technical matters are kept secret and not shared down the line, fearing that someone will leak it. The visitors are not allowed in specific area, whereas they do not make efforts to keep their staff with them. When people are leaving the company and going out, how can a secret be maintained? Further when that senior man leaves, no one shall know the method of managing. If they share the technical matters down the line, the people down the line shall be able to manage the show even in the absence of senior person.

Case No. 24 – Customer requirement and quality assurance

In almost all the mills or garment factories, there are checkers and quality control people at all levels who check the material and reject if any variation is found. They want to ensure that customer does not get bad quality material. A spinner takes all care to ensure that different lots of cotton or yarn are not mixed; whereas in knitting and in weaving, the yarns left out from different lots are mixed to get the realization. In processing, the quality assurance people try their level best and if there is variation in shade, they reject the materials; whereas in fashion industry, mixing panels of different depths of shade is presented as a new style and accepted by the ultimate customers also.

A very interesting thing noticed is in retail marketing. In a retail shop, there shall be hundreds of varieties, and the sale of a piece depends on the sales person, the products he shows, the method in which he explains the plus points of the product, the way in which he deals with the customer, and so on. What are all the points checked by quality control inspectors in spinning, weaving, or processing are never verified by the ultimate customers, whereas they are all used as profit earning points by the middle men. While shade variation is considered as a problem, the faded jeans became a fashion. The designs and motifs which were considered as ugly at one time have become a fashion now. Similarly in the garment dimensions and measurements. In the garment factory, if there is no symmetry, the garment is rejected; whereas in fashion, you can see one side up and other side down. Intelligent merchants are selling the so called defective pieces in different trade names and claim them as "Self-Design", "Only one piece available on earth in this design", "Unique and innovative, never in the history", and so on, and are getting very high premiums at high end market.

Case No. 25 – Bringing efficient staff from a good mill

The managing director of a mill was not happy with the working style of his chief executive and felt that the company was running under loss because of him, whereas his competitor mill was doing better. The managing director somehow developed contact with the CEO of the competitor mills, offers a package of 50% more than what he was drawing there, and requested to join his company. The moment new CEO came, he wanted the complete technical team to be changed and offered to bring the whole team from his earlier company. All his team members from his earlier company were offered 25–35% higher packages and were brought to this mill with lucrative designations. Suddenly the competitor mill had to face the shortage of staff and they absorbed all the staff members who were made to leave this mill, and offered 20%–30% more for all to ensure that they do not leave and go off. The performance of the mill did not improve as these were not the people responsible for that. It was the management policy and the strategic decisions taken by the managing director of the competitor company who guided the company to run under profit.

Case No. 26 – Absenteeism due to marriage at relative's place

A large textile mill in north Karnataka, situated in a rural area, had its workforce settled in the company's quarters and nearby villages. The mill being old, the workmen also were there from generations and had developed relations among themselves. Whenever there was a marriage or any function at any house, the absenteeism used to be very high. Normally the absenteeism was very high between 15th February to 15th June every year due to marriages, *jathras*, and holidays in the schools. The management discussed the problem with the Swamiji of local Mutt and negotiated with workmen and came out with a novel idea of mass marriages. The mills took responsibility of providing food for the marriage; whereas a nominal fee was collected from the parties getting married for the materials and pundit for marriage. The Swamiji shall be present in the marriage and bless each couple. It became very economical for the workers. All the marriages fixed in a year were held on *Aksha Thritiya*, an auspicious day. By this, the production did not suffer on the other days due to mass absenteeism.

In the end

There are millions of such case studies available in textile industry because of its vast spread, thousands of years of experience, varying culture between regions and between people, different level of knowledge and technology available, and so on. However, one with cool approach and positive attitude of working with people as a team member can overcome the problems or learn to live successfully with the problem and see that industry moves forward.

Learning to live with the problem is very important where we are not the party to take action to eliminate the root of the problem. We have our own boundary of working and would be successful if we can somehow manage the situation and move forward. When we recollect our experiences, it looks really wonderful; and hence it is referred as "Wonderland of Problems". One who is daring and takes quick decisions is liked by the people and the one who is very systematic, trying to find the root cause of the problem by scientific methods are termed as unpractical and rejected.

Only studying or understanding the facts cannot help in solving problems and achieving the results. A number of problems are not solved in textile and garment industry because people wait for someone to come forward and take action. If you need to solve your problem, you need to move forward and take action. There is a *subhashita* in Sanskrit which is as follows:

उध्यमेन हि सिध्यन्ति कार्याणि न मनोरथैः

न हि सुप्तस्य सिंहस्य प्रविशन्ति मुखे मृगः

Any work will not get accomplished just merely by desiring for its completion. A 'prey' by itself doesn't enter in to the sleeping lion's mouth!!!

References

- 5 Whys – Wikipedia
- A Few Tools for Continuous Improvement – http://www.mavtechglobal.com
- A Teacher in Taoyuan – http://hallhouston.blogspot.in
- Action Centered Leadership – http://www.businessballs.com/action.htm
- Advertisement in internet regarding ISO 9001 certificates viz. www.nimbuscertifications.com/9892413039, www.iqmindia.com/9001.html , www.isocatalyst.com
- Arnaud – Be MECE (Mutually Exclusive and Collective Exhaustive) – http://powerful-problem-solving.com
- Arnaud – Use Process Appropriately – http://powerful-problem-solving.com/use-processes-appropriately
- Basics of Quality Leadership – By Tata Quality Management Services.
- Bednarz, T. F. The Six Phases of Critical Thinking – http://blog.majoriumbusinesspress.com
- Boost your creative problem solving skills – Become a highly creative problem solver and brainstormer – http://www.hypnosisdownloads.com/thinking-skills/problem-solving
- Boundary Analysis – http://www.volusia.org/gis/boundary.htm
- Brainstorming – Generating many radical creative ideas – http://www.mindtools.com/brainstm.html
- Brainstorming – http://www.engin.umich.edu/~problemsolving
- Brainstorming – http://www.nwlink.com/~donclark/perform/brainstorm.html
- Brainstorming – Wikipedia – http://en.wikipedia.org/wiki/Brainstorming
- Brainstorming Software – Wikipedia – http://en.wikipedia.org/wiki/Brainstorming_software

- Brainstorming, Buzz Group, and Snow Balling – http://teachingtechniques.boston.ac.uk/brainstorm_buzz_groups_and_snow_balling.html

- Brainwriting – http://www.mycoted.com/Brainwriting

- Bullet Proofing – http://www.mycoted.com/BulletProofing

- By Harry Joiner – Seven Steps to Problem Solving– Forums, Logistics Today.

- Causal Mapping – http://pictureitsolved.com/resources/practices/causal-mapping

- Change management – http://www.businessballs.com/changemanagement.htm

- Chapman, A. Change Management – Business Balls.com

- Check Sheets – http://en.wikipedia.org/wiki/Check_sheet

- Checklist – http://en.wikipedia.org/wiki/Checklist

- Clayton, J. The Five Stages of the Strategic Management Process, Demand Media – http://smallbusiness.chron.com

- Cochran, C. Six Problem Solving Fundamentals – Quality Digest

- Control chart – http://en.wikipedia.org/wiki/Control_chart

- Creative Problem Solving – http://www2.hawaii.edu/suremath/home.html

- Creative Problem Solving – Resources for practitioners by Omniskills LLC – http://www.creativeproblemsolving.com

- Creativity and Innovation Techniques – an A to Z – http://www.mycoted.com/Category:Creativity_Techniques

- Critical Thinking – http://en.wikipedia.org/wiki/Critical_thinking

- Culture as culprit: Four Steps to Effective Change – http://executiveeducation.wharton.upenn.edu

- Data Collection – http://en.wikipedia.org

- Davidmann, M. Directing and Managing Change.

- Decision Tree – http://en.wikipedia.org

- Decision Trees – http://www.mindtools.com/dectree.html

- Decisional Balance Sheet – Wikipedia

- Defect Concentration Diagram – http://en.wikipedia.org

- Delphi Method – http://www.rand.org/topics/delphi-method.html

- Diagnose – http://dictionary.reference.com

- Dictionary.com – dictionary.reference.com/browse/serendipity

- Dr. Devaraja, T. S. Indian Textile and Garment Industry–An Overview by, http-sibresearch.org-uploads-2-7-9-9-2799227 - Department of Commerce, Post Graduate Centre, University of Mysore, Hassan, India 2011

- Dr. Joel. R. DeLuca. Evoked Sidebands – An adventure in building understanding – by Overcoming Resistance to Change

- Dr. Pio, E. and Natarajan, T. R. Quality Samurai

- Dr. Wenger, W. Beyond Methods – 20 Points to help you solve problems – www.winwenger.com/20points.htm

- Dr. Wenger, W. Double-Entry A-ha! Method

- Dr. Wenger, W. Gravel Gulch – 4 Steps to Problem Solving.

- Dr. Wenger, W. Over-the-Wall, *Basic Post-Einsteinian Discovery Technique for Creative Solution-Finding.*

- Dr. Wenger, W. Taxonomy of Methods – A Partial Summary of All Possible Techniques for Solving Problems.

- *Edward De Bono – Lateral thinking: creativity step by step* – http://homeworktips.about.com/od/brainchallenges/a/Lateral-Thinking.htm

- Effective Problem Solving – University of South Australia

- Excursions – A Mental Flight of Fancy – http://www.creativerealities.com

- Exforsys Inc – The Connection Between Innovation and Problem Solving – http://www.exforsys.com

- Exforsys Inc Creative Problem Solving – http://www.exforsys.com/career-center/problem-solving/creative-problem-solving.html

- Exforsys.Inc – How Brainstorming Can Help You Solve Problems – http://www.exforsys.com

- Exforsys.Inc – How Critical Thinking Can Help You Solve Problems - http://www.exforsys.com

- Exforsys.Inc – How Role Playing Can Help You Solve Problems – http://www.exforsys.com

- Exforsys.Inc – How To Effectively Solve Problems – http://www.exforsys.com

- Exforsys.Inc – How To Properly Approach a Problem – http://www.exforsys.com

- Exforsys.Inc – Problem Solving Strategies – http://www.exforsys.com

- Exforsys.Inc – The Importance of Reductionism In The Problem Solving Process – http://www.exforsys.com

- Exforsys.Inc – The Use of Trial and Error To Solve Problems – http://www.exforsys.com

- Exforsys.Inc – Wicked Problems – http://www.exforsys.com

- Famous Models – Adair's Three Circles – http://www.chimaeraconsulting.com

- Firestien, R. (1989). Why Didn't I Think of That? A Personal and Professional Guide to Better Ideas and Decision Making.

- Flow Chart – http://en.wikipedia.org

- Fogler, H. S. and LeBlanc, S. E. (1995). First Steps in Solving Open-Ended Problems from *Strategies for Creative Problem Solving*.

- Folger and LeBlanc (1995). Strategies for Creative Problem Solving, Prentice Hall PTR, Englewood Cliffs, New Jersey.

- Foursight consulting – How to evaluate ideas – www.foursightconsulting.com

- Free association – http://www.thefreedictionary.com

- Free association (psychology) – http://en.wikipedia.org/wiki/Free_association_(psychology)

- Friend, J. and Hickling, A. (1987). Planning Under Pressure: The Strategic Choice Approach.

- Grid analysis – Making decision by weighing different factors – http://www.mindtools.com

- Hicks, T. (1999). Here are Seven Steps to Effective Problem-solving, Pacific Business News (Honolulu).

- Histogram – http://en.wikipedia.org

- History of Clothing and Textiles – http://en.wikipedia.org

- How to develop and demonstrate your problem-solving skills – http://www.kent.ac.uk

- How to introduce and use creative thinking techniques in brainstorming sessions – http://www.brainstorming.co.uk

- Hutton, D. W. The Change Agents Handbook

- Hypnosis – http://en.wikipedia.org

- Idea Evolution Methods and Techniques – *Creative trainer II,* www. creative-trainer.eu

- Individual brainstorming – Mind tools – http://www.mindtools.com

- Ishikawa diagram – http://en.wikipedia.org

- Jayalaxmi J. Shah – A Comparative Analysis of Two Major Cotton Textile Centres of India-Bombay and Ahmadabad, Department of Geography SNDT Arts and SCB Commerce College for Women-Bombay - *www.raco.cat-index.php-treballsscgeografia*

- Joseph A. Montagna. The Industrial Revolution – http://www.yale.edu

- Lao-tzu Quotes – The Quotations Page

- Leclerc, O. and Moldoveanu, M. Five routes to more innovative problem solving – http://www.mckinsey.com

- Linked Training Experts – Are you storming or drizzling – How to make brainstorming work – http://www.linkedtraining.com

- Listing Pros and Cons – http://www.mycoted.com

- Lyndsay Swinton. Seven Brainstorming Rules and Techniques to Get More from Group Problem Solving, Management for the Rest of Us – www.mftrou.com

- Manage Train Lean – Innovation Techniques: SCAMMPERR – http:// www.managetrainlearn.com

- Management – Formal Planning Techniques – http://www.docstoc.com

- Management Techniques: Force Field Analysis – http://www.odi.org. uk

- Metaphorical Thinking – http://www.zideate.com

- Mind Mapping – Wikipedia – https://en.wikipedia.org

- Mindtools.com

- NAF – http:// www.mycoated.com

- Negative Brainstorming – http://www.mycoted.com

- Neuro-linguistic programming – Wikipedia – http://en.wikipedia.org

- Normal Group Technique – Business Dictionary.com – http://www. businessdictionary.com

- Observer and Merged View Points – http://www.mycoted.com

- Osborn and Alex, F. (1957). Applied Imagination: Principles and Procedures of Creative Problem-solving, New York: Scribner
- Osborn, A. F. Father of the Brainstorm – http://www.skymark.com/resources/leaders/osborne.asp
- Other Peoples Definitions – http://www.mycoted.com
- Other Peoples Viewpoints – http://www.mycoted.com
- Paired Comparison Analysis – http://www.mindtools.com
- Pair-wise Comparison – http://en.wikipedia.org
- Panel Consensus – http://www.mycoated.com
- Pareto Analysis – http://en.wikipedia.org
- Personal Balance Sheet – Mycoted – www.mycoted.com
- Phases of Integrated Problem Solving – Mycoated
- Potential Problem Analysis – http://www.mycoted.com
- Potential Problem Analysis – Mycoted.com
- Powerful Problem Solving – Diagnose the problem – http://powerful-problem-solving.com
- Preliminary Questions – Mycoted – www.mycoted.com
- Press Information Bureau, Government of India, Ministry of Textiles, 14-May-2012 15:43 IST, Hank Yarn Obligation – http://pib.nic.in
- Principle Component Analysis – https://en.wikipedia.org
- Problem Centered Leadership – http://www.mycoted.com
- Process Mapping – http://www.fpm.iastate.edu
- Pulses, Potentials, and Concerns – Mycoted.com
- Purushothama, B. (2006). Process Management in Textiles, CVG Books.
- Purushothama, B. (2007). Five Golden Questions – A self-assessment Tool – Quality Update. Indian Society for Quality.
- Purushothama, B. (2007). Winning Strategies – Pubadchi Publications.
- Purushothama, B. and Srinivasu, N. (1994). Total Quality Management, Gokak Mills Internal Publication
- Purushothama, B. Linking Exercises, Quality Update Aug 2007 – Indian Society for Quality.
- Pushing through the problem – By Small Group Communications

- Random Stimuli – http://www.mycoated.com

- Rawlinson, J. G. (1981). Creative Thinking and Brainstorming.

- Rawlinson, J. W. Introduction to Creative Thinking and Brainstorming, British Institute of Management.

- Rawlison Brainstorming – http://www.mycoted.com

- Receptivity to Ideas – http://www.mycoated.com

- Receptivity to Ideas – http://www.mycoted.com

- Remedy – http://www.thefreedictionary.com

- Reverse Brainstorming – A Different Approach to Brainstorming – http://www.mindtools.com

- Role Storming – http://www.innotour.com

- Role Storming – http://www.mycoted.com

- Ronold, J. Questorming – http://pynthan.com

- RPR Problem Diagnosis – http://en.wikipedia.org

- Run Chart – http://en.wikipedia.org

- Sama dana bheda danda – http://slowtrain1.blogspot.in

- Sama, Dama, Bheada, Danda – http://en.wikipedia.org

- Samskruta Mouktikaani – http://sanskritpearls.blogspot.in

- Scammer – www.merriam-webster.com

- SCAMMPERR – http://www.mycoted.com

- SCAMPER – http://www.mycoated.com

- Scatter Plot – http://en.wikipedia.org

- Scatter Plot – http://en.wikipedia.org

- Scheffer, B. K. and M. G. Rubenfeld (2000). A Consensus Statement on Critical Thinking in Nursing, *Journal of Nursing Education*, **39**, pp. 352–359.

- Scheffer, B. K. and Rubenfeld, M. G. Critical Thinking: What is it and How do We Teach It? Current issues in Nursing, Grace, J. M. and Rubl, H. K. (2001)

- Scott G Isaksen – Human Factors for Innovative Problem Solving – http://www. cpsh.com

- SDI Systematized Direct Induction – http://www.mycoted.com

- Search Conference – http://www.mycoated.com

- Seven basic Tools of Quality – http://en.wikipedia.org

- Seven Steps for Problem Solving – http://www.pitt.edu/~groups/probsolv.html

- Shibata, H. (1997, 1998). Problem Solving: *Definition, Terminology, and Patterns,* Copy rights © H. Shibata all reserved.

- Shyam Talwadekar – Visual Management Through Five S: A Japanese Tool of Kaizen."

- Similarities and Differences – http://www.mycoated.com

- Simple Rating Methods – http://www.mycoted.com

- Simplex – http://www.mycoated.com

- Slice and Dice – http://whatis.techtarget.com

- Slice and Dice – http://www.mycoated.com

- Socratic Method – http://en.wikipedia.org

- Sri Ravishankar – Love and Sin – http://srisriravishankar.org

- Stakeholder Analysis – Winning Support for your projects – http://www.mindtools.com

- Stakeholder Analysis Worksheet – www.creativeproblemsolving.com

- Stephen Brewster – Me – Do you SCAMMPERR? – http://stephenbrewster.me

- Strategic assumption surfacing and testing – http://en.wikipedia.org

- Strategic assumption surfacing and testing: integrating world views – http://epress.anu.edu.au

- Strategic Assumption Testing – http://www.mycoted.com

- Strategic Choice Approach – http://www.mycoted.com

- Strategic Management Process – http://www.mycoted.com

- Strategic Options Development and Analysis – http://www.mycoated. SODA

- Stratification – http://asq.org

- Subhashitani – http://s3.amazonaws.com

- Successive Element Integration – http://www.mycoted.com

- Sue Dinwiddie – Kid Source On line – "I want my way" Problem-Solving Techniques with Children Two to Eight

- Swamy Krishnanand – The Spiritual Import of the Mahabharata and the Bhagavadgita – http://www.swami-krishnananda.org

- SWOT Analysis – http://www.mycoated.com

- Sylvan, P. Creativity, Innovation, and Problem Solving, Quantum Books.

- Synectics – http://en.wikipedia.org

- Synectics – http://www.mycoted.com

- The Culture of Safety – American Society of Safety Engineers – www. asse.org

- Thoburn, J. The Impact of World Recession on the Textile and Garment Industries of Asia – www.unido.org-fileadmin-user_media-publication.pdf

- Toffler, A. (1971). The Future Shock

- Using Stakeholder's Analysis as a Problem Solving Tool – http://peoplepower73.hubpages.com

- UT4 Application Report – Uster Technologies – www.uster.com

- VanGundy (1988). Techniques of Structured Problem Solving.

- What is the meaning of "saam, daam, dand, bhed"? – www.krishna.com

- Wikipedia Encyclopedia